Large-area Effects of GM-Crop Cultivation

Theorie in der Ökologie

Herausgegeben von Broder Breckling

Band 16

PETER LANG
Frankfurt am Main · Berlin · Bern · Bruxelles · New York · Oxford · Wien

Broder Breckling
Richard Verhoeven
(eds.)

Large-area Effects of GM-Crop Cultivation

Proceedings of the Second
GMLS-Conference 2010 in Bremen

PETER LANG
Internationaler Verlag der Wissenschaften

Bibliographic Information published by the Deutsche Nationalbibliothek
The Deutsche Nationalbibliothek lists this publication in the Deutsche
Nationalbibliografie; detailed bibliographic data is available in the internet at
http://dnb.d-nb.de.

Cover illustration:
© Broder Breckling

Typesetting & Lay-out:
Richard Verhoeven

ISSN 1615-374X
ISBN 978-3-631-60361-1
© Peter Lang GmbH
Internationaler Verlag der Wissenschaften
Frankfurt am Main 2010
All rights reserved.

All parts of this publication are protected by copyright. Any
utilisation outside the strict limits of the copyright law, without
the permission of the publisher, is forbidden and liable to
prosecution. This applies in particular to reproductions,
translations, microfilming, and storage and processing in
electronic retrieval systems.

www.peterlang.de

Contents

Foreword   9

Organising committee and reviewers   10

Welcome Address Karin Mathes, Vice President of the Bremen State Parliament   11

Welcome Address of Beatrix Tappeser, Head of GMO Regulation/Biosafety, Federal Agency for Nature Conservation, Germany   14

Welcome Address of Thomas S. Hoffmeister, Vice Dean Elect of the Faculty Biology / Chemistry, University of Bremen   17

Chapter I: Risk assessment and monitoring of GMO-effects

Exposure of maize harvest by-products to aquatic ecosystems and protected nature reserves in Brandenburg, Germany
    Werner Kratz, Christina Mante, Frieder Hofmann, Ulrich Schlechtriemen, Ulrike Kuhn, Steffi Ober & Rudolph Vögel   21

Networks and environmental observation programs as a tool for general surveillance – first experience and future requirements
    Wiebke Züghart   24

Genetically modified crop consumption at large scale: Possible negative health impacts due to holes in assessment. Overview of the safety studies of GMOs performed on mammals
    Gilles-Eric Séralini, Joël Spiroux de Vendomois, Dominique Cellier, Robin Mesnage & Emilie Clair   28

Roundup® in genetically modified plants: Regulation and toxicity in mammals
    Robin Mesnage, Emilie Clair & Gilles-Eric Séralini   31

Seedling emergence of oilseed rape (B. napus L.) and wild relatives on ruderal soils
    Jana Seeger, Broder Breckling & Juliane Filser   34

Setup, efforts and significance of a GMO monitoring program – An Austrian case study
    Kathrin Pascher, Dietmar Moser, Stefan Dullinger, Leopold Sachslehner, Patrick Gros, Norbert Sauberer, Andreas Traxler, Georg Grabherr & Thomas Frank   37

Prioritizing GMO monitoring sites in dynamic cultivation systems and their environment – a conceptual on-farm approach
    Claudia Bethwell, Frieder Graef, Ulrich Stachow, Hans-Jürgen Müller & Frank Eulenstein   39

Chapter II: Dispersal and coexistence of GMO

Potential GM-maize cropping in Schleswig-Holstein I: Spatial heterogeneity of GM cultivation (Scenarios)
    Christiane Eschenbach, Wilhelm Windhorst & Andreas Rinker   47

Potential GM-maize cropping in Schleswig-Holstein II: Model and GIS based approaches to estimate the GM-share in conventional maize yield
    Christiane Eschenbach, Broder Breckling, Andreas Rinker & Wilhelm Windhorst   51

WebGIS as a tool for GMO monitoring support and for identification of potential coexistence problems due to GMO cultivation
    Lukas Kleppin, Gunther Schmidt & Winfried Schröder   56

An African perspective of GM maize gene flow
    Chris Viljoen & Lukeshni Chetty   60

Proposal for large-scale regional monitoring of genetically modified maize crops in small-scale agricultural systems in Africa
    Denis Worlanyo Aheto   62

Influence of GM-crop cultivation on local apiculture and floral environment
    Peter Wagner   65

Effect of two different gap crops on pollen-mediated gene flow in maize
    Maren Langhof, Bernd Hommel, Alexandra Hüsken, Aldona Jarzmik, Joachim Schiemann, Peter Wehling & Gerhard Rühl   68

Chapter III: Management and control of GM-crops at large spatial scales

Comparative ecological effects of GM and other innovations in maritime arable-grass production systems
    Geoffrey Squire, Mark Young & Alison Karley — 73

Demography of feral oilseed rape over 11 years in an agricultural region
    Gillian Banks, Mark W. Young & Geoffrey R. Squire — 76

The Triffid case: A short résumé on the re-discovery of a de-registered GMO
    Gunther Schmidt & Broder Breckling — 79

Monitoring maize diversity in Mexico for decision making
    Francisca Acevedo Gasman — 82

Study of maize fields and their surroundings in European regions regarding the suitability for coexistence of different maize cultivars
    Sabine Prescher, Joachim Schiemann & Alexandra Hüsken — 84

Is there any room for alternatives? Socio-economic implications of GMOs cultivation at large-scale – Case study in Spain
    Rosa Binimelis, Iliana Monterroso & Mariel Vilella — 89

Development of the indicator "Genetic engineering in agriculture"
    Christiane Eschenbach & Wilhelm Windhorst — 91

New pest in crop caused by large scale cultivation of Bt corn
    Christoph Then — 94

Chapter IV: Setting the frames: integrative interdisciplinary approaches

The Danish coexistence regulation and the Danish farmers attitude towards GMO
    Morten Gylling — 101

From risk assessment to in-context trajectory evaluation: GMOs and their social implications
    Vincenzo Pavone, Joanna Goven & Riccardo Guarino — 105

The lack of regulation on GMO as one of the risk factors for biodiversity in a place of unique value – Example of the Lake Baikal Region
    Natalia Sirina, Christiane Eschenbach, Wilhelm Windhorst & Felix Müller — 109

GM maize and oil seed rape in Germany: Economic welfare losses from large scale adoption scenarios
    Jan Barkmann, Manuel Thiel, Ludwig Theuvsen, Christiane Eschenbach, Wilhelm Windhorst & Rainer Marggraf   114

Risk Governance: Communication Strategies for Coexistence with GMOs (Genetically Modified Organisms)
    Claudia Bethwell, Thomas Weith & Klaus Müller   117

EU and German law on coexistence: Individual and systemic solutions and their compatibility with property rights
    Sarah Stoppe-Ramadan & Gerd Winter   121

Legal implications of the step-by-step principle on risk assessment of GMOs
    Caroline von Kries & Gerd Winter   125

Dead end developments – lessons learned from unsuccessful GMO
    Broder Breckling   129

Foreword

After the great success of the first GMLS-conference in spring 2008, we were encouraged to continue the process of discussion on ecological effects of Genetically Modified Organisms (GMO) on the regional extent. This volume presents the proceedings of the second conference on "Implications of GM-Crop Cultivation at Large Spatial Scales". Again, we invited international experts from science, administration and jurisprudence to the University of Bremen, this time from March 25–26, 2010.

The conference provided a platform to discuss ecological, agricultural and economic implications of GM-cultivation. The proceedings present new developments in risk assessment and monitoring of GMO and new results in scientific modelling. They give insight into diverse approaches for co-existence regulations or nature protection standards in different countries inside and outside the European Union. The reader will also find contributions on legal developments in the European Union as well as specific regulations in some single countries.

In the second conference, we extended the discussion to cover additional perspectives, following a fruitful approach to invite also experts from countries outside Europe. They report on the special situation of GMO-farming in their countries and the associated problems in Ghana, in South Africa as well as in Mexico, the home country of maize.

In four chapters this book documents thirty contributions to the conference which have been peer reviewed. The editors wish to cordially thank all conference delegates for their contributions as well as the German Ministry for Education and Research (BMBF) for the funding of the conference within the Social Ecological Research programme.

Broder Breckling, Richard Verhoeven

Organising committee of the conference

Broder Breckling
(University of Bremen, University of Vechta)

Wolfgang Büchs
(German Ecological Society, SG on Agroecology)

Christian Eschenbach
(University of Kiel)

Florian Keil
(Institute of Social Ecological Research, Frankfurt am Main)

Hartmut Meyer
(German Ecological Society, SG Genetic Engineering and Ecology)

Winfried Schröder
University of Vechta

Richard Verhoeven
University of Bremen

Wiebke Züghart
(Federal Agency for Nature Conservation, Germany)

Karin Mathes
Vice President of the Bremen State Parliament

**Ladies and Gentlemen, dear guests,
Welcome to the Free Hanseatic City of Bremen**

As vice president of the Bremen State Parliament it is a pleasure for me to welcome you to the GMLS Conference on Implications of Genetically Modified Crop Cultivation at Large Spatial Scales here at the University of Bremen. It is the second time that Bremen hosts this conference. The Conference attempts to bring together leading scientific expertise to assess impacts of genetically modified organisms in the context of agricultural applications. Let me first say some words why it is a good choice to locate this event here. Afterwards I will comment the relevance of the conference topic from a political perspective.

Bremen – the two city state in Northern Germany

The Free Hanseatic City of Bremen is the working and living environment for about a half million of inhabitants, and it is a historical place. It has seen centuries of growth, of fundamental changes, new opportunities but also threats and disasters. Looking back in history, it is apparent that political leadership has to encourage the conditions of accessible gains and the sharing of benefits. Also, the responsibility of politics is to restrict and regulate risks in a way that hazard and damage can be avoided and managed respectively.

Starting in the 7th century as a medieval trading location, the economic, technical and cultural life of the city always had to manage risks. In the middle ages, Bremen as a part of the Hanse, brought the fist insurance systems on the basis of mutual support to defend from economic risk. Later, Bremen was involved in industrialisation, trade and overseas exchange. The city survived all national re-organisations as an independent political body in the federal structure of Germany. It is now the smallest German federal state, consisting of the two cities of Bremen and Bremerhaven. Today, it plays an important role in space technology, car and steel manufacturing and wind energy. It is also a location of important scientific institutions like the Alfred Wegener Institute of Polar Research in Bremerhaven.

Bremen's long experience in risk assessment

For the public institutions as well as the private sector a good considered risk assessment is indispensable. This applies to genetically modified organisms and to other technical approaches. Having worked for many years as a senior research scientist in general ecology and ecotoxicology, I am well familiar with the requirements of environmental risk assessment. Compared to chemicals, GMO require a more comprehensive analysis. This encompasses not only a compositional analysis but also assessment of physiological performance, aspects of cultivation and ecological effects as changes in biodiversity. Monitoring is an additional task. Which aspects have to be considered are currently debated by the EFSA.

To make reasonable decisions, policy depends on an information basis that is well balanced. It is highly important, that not only the view of the notifiers is publicly available. For a reasonable risk assessment, critical and independent research is indispensable. The public funding of scientific expertise that is not involved in a specific interest is a MUST for regulators if they want to be efficient.

To improve risk assessment considerable tasks have to be solved. Advanced metabolic methods exist which can be used to compare the cellular biochemistry of the GMO with an appropriate conventional comparator. Deficits in risk research exist concerning the characterisation of the receiving environment. This comprises the agricultural structures as well as biodiversity implications of GMO cultivation. It is a primary task of research and funding policy to assure the capacities of independent analysis.

I expect that the conference results and its documentation will be appropriately considered on the scientific, on the public and on the administrative level. For this conference, the organising committee has brought together contributions from Europe and overseas which provide important new insight and experiences. So I expect a relevant impact in the discussion on GMO and the regulation of the involved risks.

GMO policy in Bremen

Let me inform you about some political decisions the Free Hanseatic City of Bremen has made. Bremen has decided that agricultural contractual partners of the city working on municipal areas are obliged to cultivate conventional varieties. Bremen also encourages farmers to establish private contracts to manage their farms GMO free. In the general public, this policy has a considerable support. As a politician, I have to emphasise, that in decision making on GMO, scientific information is highly important. Value-based consumer preferences and the protection of GMO free production are at least equally relevant. The value preference for food being as natural as possible has a very high priority in consumer decisions.

Enjoy your stay in Bremen

I hope, that your stay in Bremen offers useful scientific information, and hopefully even more than that. You should take some time to enjoy the highlights in the city. The Dome with its medieval Roman Style architecture is certainly worth to be visited, as well the Renaissance Townhall, a UNESCO world heritage site. The Ratskeller in the basement of the townhall offers a large collection of German wines. You might also like to walk along the River Weser with its diversity of pubs and restaurants offering a wide variety of international cuisine. Those of you who are culturally interested might like to visit the Schnoor as the most ancient medieval part of the city. You will find additional inspiration in the Overseas Museum right next to the main station, or in the various Arts Galleries.

I wish you an enjoyable stay in Bremen and that you establish lasting contacts. Exchange with our scientific institutions might bring further inspirations for your personal work.

Bremen, March 25, 2010
Karin Mathes

Beatrix Tappeser
Head of GMO Regulation/Biosafety
Federal Agency for Nature Conservation, Germany

Dear colleagues,

I have the honour to open this conference on behalf of the Federal Agency for Nature Conservation and like to welcome you. The Federal Agency is one of the competent authorities in Germany for the approval of GMO. As in Europe we have also a complex regulatory system in Germany involving federal agencies and state authorities. Our lead authority is the "Office of Consumer Protection and Food Safety" who issues the final statements. The special task of the Federal Agency for Nature Conservation is the Environmental Risk Assessment of GMO and the evaluation of monitoring plans and conceptual work on GMO monitoring.

There are a number of obligations and recommendations when conducting an Environmental Risk Assessment informed by international law, EU-wide law and German law: case-by-case, step-by-step, taking into account direct and indirect, immediate and delayed as well as cumulative long-term effects. And this shall be done in accordance with the precautionary principle.

It is not an easy task to translate these obligations and recommendations into concrete requirements for implementation. In addition there are some overarching issues as for example obligations deriving from decisions of the Convention for Biodiversity normally not explicitly addressed when conducting a RA. But for the benefit of coherence of international, EU-wide and national decisions and policy goals these decisions should form the background for the implementation.

„A safe operating space for humanity" is the title of a comprehensive report published by leading climate change scientists worldwide together with the Stockholm Resilience Centre. A short summary can be found in the issue of September 24, 2009 of the journal *Nature*. Within this report the authors tried to identify and summarize Earth system processes and associated thresholds which, if crossed, could generate unacceptable environmental damage. They define nine such processes, besides climate change, interference with nitrogen and phosphorus cycles, stratospheric ozone depletion, ocean acidification, global freshwater use, change in land use, chemical pollution, atmospheric aerosol loading and the rate of biodiversity loss. According to the authors the boundaries of three systems have already been exceeded. One of these is the rate of biodiversity loss. Due to their analysis the rate of species distinction has accelerated in such a way that it is 100 to 1000 times what could be considered natural. Main drivers are considered to be industrialized forms of agriculture and changes in land use.

2010 is the International Year of Biodiversity. During the official opening on the 11th of January our chancellor Angela Merkel acknowledged that "…the question of conservation of biodiversity is of the same dimension and importance as the mitigation of climate change…"

As pointed out agriculture is identified as one of the main drivers of biodiversity loss. Since the middle of the last century the rate of species distinction is huge due to intensification, rationalisation, specialisation and a depletion of landscape structures. It seems obvious that such goals like "Stop the loss" as adopted by the Conference of the Parties to the Convention on Biodiversity can only be achieved when changes and improvements in agricultural management and practice will be implemented.

Worldwide the understanding of an Environmental Risk Assessment (ERA) of Genetically Modified Plants (GMP) is as such that it should be based on a comparative approach. Only when the impact of a given GMP is worse than the normal agricultural practice with its conventional counterpart then this impact may be characterised as an adverse effect. That's the mainstream understanding. But what does normal agricultural practice mean? Farms in southern Germany are quite different from farms in the northeast, climate, soil and landscape structure are also different. And what is relevant and challenging for Germany is even more relevant and challenging when it comes to EU wide approvals.

The draft guidance document on ERA just opened for public comment sees the management systems as important parts of the characterisation of the receiving environment together with the geographical zones and the GMP itself. But again, which management systems are deemed relevant in this context?

Hence definition and characterisation of the adequate comparison (with the concurrent biodiversity impact) is of utmost importance. Just following the status quo seems to be not an option at least when the aims of the CBD are taken seriously.

Agriculture in Germany is closely interlinked with non managed habitats, forests, water streams and protected areas with different protection status besides of infrastructure constructions and urban structures. When following the wording of Annex II about Principles of the ERA of Directive 2001/18/EC it is recommended that "…an analysis of the cumulative long-term effects relevant to the release and the placing on the market is to be carried out." To fulfil this requirement there is a need of the integration of small scale data collections into models which allow scaling up. This translates in the need for landscape level data. And for the future hopefully reliable science based monitoring data for validation of the developed models are also available. With that we are in the middle of the topics of this conference.

When looking into the development pipelines and the regulatory pipeline there is an additional need to model the possible impacts of different plants with the same engineered traits and the same plants engineered to express different traits.

And to make it even more complex on the long run we need models which integrate the cultivation of annual plants together with perennial plants having the same and different traits.

For example there are a number of additional herbicide resistant traits in the regulatory or development pipeline. In addition to glyphosate and glufosinate resistant plants industry is working on Dicamba, 2,4 D, Acetolactate synthase (ALS), Sulfonylurea, and Isozaflutole resistant plants as well as plants containing hydroxyphenyl pyruvate dioxygenease inhibitors (HPPD inhibitors), all rendering the plants herbicide resistant.[1] Stacking of these genes will increase in the next years. Given the results of the Farm Scale Evaluations in Great Britain published 2005[2] that even one year of cultivation of single herbicide-resistant trait plants may have biodiversity impacts the question arises what may be the impact of the parallel use of a number of broad spectrum herbicides in consecutive years.

Therefore this second conference on "Implications of GM-Crop Cultivation at Large Spatial Scales" is very timely and addresses important aspects not much discussed until now. I am looking forward to the presentations and discussions and do wish all of us an inspiring conference.

Bremen, March 25, 2010
Beatrix Tappeser

---

1 Stein A.J., Rodriguez-Cerezo E. (2009) The global pipeline of new GM crops. JRC report. EUR 23846 EN.
2 Firbank L.G., Rothery P., May M.J., Clark S.J., Scott R.J., Stuart R.C., Boffey C.W.H., Brooks D.R., Champion G.T., Haughton A.J., Hawes C., Heard M.S., Dewar A.M., Perry J.N. & Squire G.R. (2005) Effects of genetically modified herbicide-tolerant cropping systems on weed seedbanks in two years of following crop. Biology Letters of the UK Royal Society. www.journals.royalsoc.ac.uk.

Thomas S. Hoffmeister
Vice Dean Elect of the Faculty Biology / Chemistry
University of Bremen

Dear colleagues, ladies and gentlemen,

As vice-dean elect of the faculty of biology and chemistry it is my pleasure to welcome you at the University of Bremen. Let me briefly introduce our institution to you that hosts the 2nd GMLS conference.

Compared to other Universities in Germany, the University of Bremen is a relatively new and recent institution. It was founded in 1971 as a reform university. At that time the hierarchical attitude of academic tradition was seriously questioned and new solutions for educational concepts were sought. Some of the ground-breaking educational concepts implemented in those early days, which became to be known as the Bremer Modell, have since become established features of modern university education all over Germany; for example interdisciplinary study and research, research-based teaching projects, orientation to practice, and responsibility towards society.

The University of Bremen was a close runner-up at the German "Excellence Initiative in 2006" and ranks among the top ten German universities in the acquisition of third-party funding. The high quality of research is not least due to the close cooperation with the numerous research institutes which have decided to locate on the University campus. This infrastructure is attracting more and more enterprises to the adjacent Technology Park, making it one of the leading high-tech locations in Germany, hosting close to 320 companies.

In Germany, education is generally in the responsibility of the federal states with the possibility to develop specific profiles. Bremen as a small federal has a comparatively limited budged. In times of decreasing tax incomes it has thus been a real financial challenge to run such a large University with 12 Faculties. The University currently has 19.000 students, and about 300 professors, with an overall annual budget of roughly 250

Mill Euros. The rate of external funds gained by University members is comparatively high and consists of about 80 million Euros annually. This shows the high level of motivation and commitment of the University members as well as the excellent research standards.

The University rests on two scientific pillars, the natural sciences with 5 faculties and the social sciences with 7 faculties. The University of Bremen was the decisive part when Bremen won the "City of Science" award in 2005 as the first city when this price was established to support the public recognition of science.

It may be of interest for you that the Faculty of Biology and Chemistry offers a variety of study traits that qualify for an assessment of an ecological impact and environmental risk analysis in general. This provides a basis for understanding the environmental effects of GMO. This topic is also on the research agenda. Over the last ten years, the faculty gained more than one million Euro of external funding for research projects to analyse environmental implications of genetically modified organisms. And, of course, Dr. Breckling has been central for this research approach at the University of Bremen.

Our faculty now offers classical zoology and botany only in the first two years of the Bachelor programme Biology and then allows for specialization in four different sub-disciplines, namely Biochemistry and Molecular Biology, Ecology, Marine Biology, and Neurosciences. Bremen has a specific profile in marine sciences, brain research, and ecology. Ecology is one of the subjects that is taught in 1 out of 6 Master's programmes of our faculty that all use English as instruction language.

The recently founded programme in Ecology is co-organised by four ecologically oriented working groups: the department for general and theoretical ecology, the department of vegetation ecology and conservation biology, my department, Population and Evolutionary Ecology, and the Botany department. The qualification covers the relevant fields of the interaction of organisms with each other and with their environment, including evolutionary strategies.

A reasonable GMO impact assessment requires a conceptual linkage of molecular processes, physiological properties, population ecology and ecosystems and landscape processes which then connect to socio-economic interactions. In this conference therefore, a spectrum of presentations ranging from biochemistry to landscape modelling and socio-economic issues is offered. Thus I hope that there is an inspiration to substantiate the necessity of bringing together these different backgrounds.

I wish you an interesting event, lots of discussion and fruitful co-operations,

Bremen, March 25, 2010
Thomas S. Hoffmeister

Chapter I

Risk assessment and monitoring of GMO-effects

Breckling, B. & Verhoeven, R. (2010) Implications of GM-Crop Cultivation at Large Spatial Scales.
Theorie in der Ökologie 16. Frankfurt, Peter Lang.

Exposure of maize harvest by-products to aquatic ecosystems and protected nature reserves in Brandenburg, Germany

Werner Kratz[a], Christina Mante[a], Frieder Hofmann[b], Ulrich Schlechtriemen[b], Ulrike Kuhn[b], Steffi Ober[c] & Rudolph Vögel[d]
([a]FU Berlin; [b]TIEM Integrierte Umweltüberwachung, Bremen; [c]NABU Berlin; [d]Landesumweltamt Brandenburg; Germany. – kratzw@zedat.fu-berlin.de)

Project background and aims

During 2005–2008, Brandenburg was one of the leading federal states in growing genetically engineered Bt-maize MON810 in Germany. Bt-maize MON810 expresses the protein Cry1Ab – originally derived from the bacterium *Bacillus thuringiensis* – which is toxic to the European corn borer. However, previous studies have also emphasized negative effects amongst others on aquatic non-target organisms (e.g. Bøhn et al. 2008; Rosi-Marshall et al. 2007), whereas Jensen et al. (2010) could not find any negative effects. Apart from insufficient knowledge on the impact level, little attention has been paid to maize pollen and harvest by-product inputs into aquatic ecosystems until today. To our knowledge, only these few studies have been published dealing with this issue so far.

Brandenburg offers plenty of aquatic ecosystems that make up to 3.4 % of its land cover. Many of them are under nature protection, e.g. by the European FFH-Directive. The aim of our project is to measure and to model the exposure of maize pollen and harvest by-products to aquatic ecosystems in Brandenburg and to evaluate appropriate measures for nature protection. Here we present preliminary results of our investigations in 2009 on deposition rates of harvest by-products in relation to the distance to the field.

Materials and methods

Study site: Harvest by-product measurements were taken on a maize field near Angermünde, Brandenburg. Formed by glacial moraines, this hilly region is one of the main maize production areas in the state Brandenburg, with annually increasing area of cultivation (Grimmert et al. 2009).

Measurements: To measure distance-related deposition rates of harvest by-products, 20 litter traps (metal pans) were placed at the ground in lee-position at a partly reaped maize field at the edge of the harvest process. For this first investigation, measurements were taken on ground and not on open water assuming no relevant differences in deposition. To test different trapping methods, traps consisted of 20 metal pans (covering 0.1 m²) placed at the ground for collecting all particle sizes. By clearing the traps after the maize harvester passed by we could measure the aerial input of by-products at defined distances (1.3, 2.7, 4.0, 5.3, 6.7, 9.3, 17.3, 25.3, 33.3, 49.3, 81.2, 127.8 m).

Samples were fractionated into three particle size classes by dry-sieving and vacuum-filtration, respectively (> 1 mm, > 63 µm and > 12 µm), dried at 60° C for 24 h and weighed. Additionally, samples were collected directly from the harvester to determine overall particle size fractions. Hereby, another two particle size classes were separated (> 125 µm and > 180 µm) and samples processed as described above. All fractions were further analysed microscopically for examining particle composition and form.

Results

Due to maize usage for silage and methane production, the whole plant is harvested and simultaneously shredded into 1–2 cm sized pieces. On a mass basis, harvest residuals are dominated by coarse particles (CP, > 1 mm), which consists of fragments of leaves, stems, cobs and kernels (Table 1). As particle size declines in the fine particle fraction (FP, < 1 mm), its gravimetric proportion to overall composition decreases markedly while the number of particles increases. Microscopically, the various plant fragments, maize fibres, pollen and starch grains can be identified in these fractions.

As shown in Figure 1 for the successively cleared litter traps, harvest by-product deposition shows a clear distance-related gradient. The total mean deposition rates of harvest residuals are about 26 g DW/m² at the edge of the maize field and decline to 0.17 g DW/m² at a distance of 133 m. Hereby, CP predominates mass inputs in the vicinity to the source (21.8 g DW/m²), but its mass proportion declines rapidly after 3–4 m. At greater distances, FP proportion predominates, since small fractions are preferentially transported by wind. In contrast to CP, deposition rates of FP declines rather constantly from about 5 g DW/m² to 3 g DW/m² at a distance of 20 m, than followed by a markedly steeper decrease down to 0.15 g DW/m² at a distance of 133 m.

Tab. 1: Proportion and main constituents of particle size fractions of maize harvest by-products (DW = dry weight).

Particle size	CP > 1 mm	FP < 1 mm total	> 180 µm	> 125 µm	> 63 µm	> 12 µm
[g] DW	40.3	3.54	3.05	0.14	0.021	0.0065
[%]	91.9	8.1	7.0	0.3	0.1	0.1
Main constituents	fragments of leaves, stems, cobs and kernels, 1–2 cm		fragments of plant tissues and elongated maize fibres	plant fragments, elongated maize fibres and grana	maize pollen, diverse plant fragments	plant fragments, grains of maize starch, fungal spores

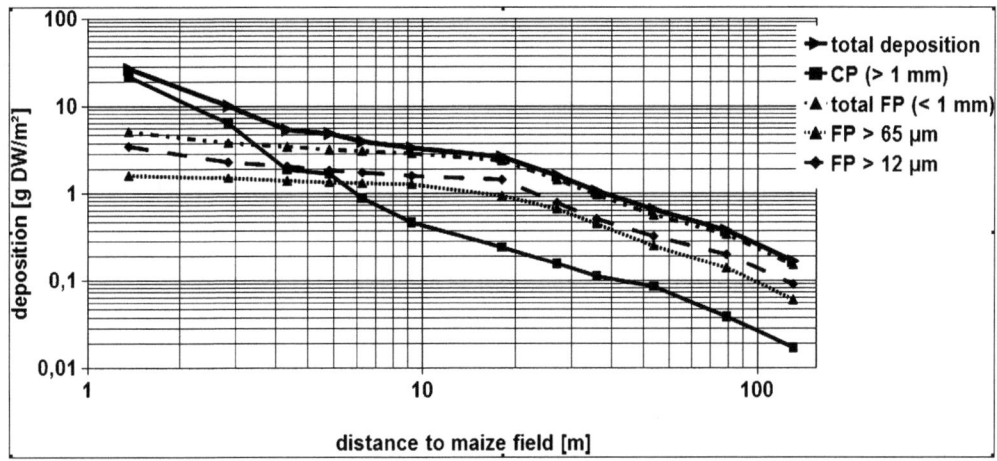

Fig. 1: Deposition of maize harvest by-products measured by metal pans (0.1 m²) in lee-position is still measurable at a distance of 100 m.

Conclusion

Input rates of maize harvest by-products show distinct gradients in relation to distance. Our results emphasize the increasing relevance of small particle fractions for deposition rates of harvest by-products at greater distances. This particle fraction has neither been evaluated in the risk assessment nor by the measurements of Rosi-Marshall et al. (2007) nor Jensen et al. (2010) and its importance has been underestimated until today. Especially, small particles have a greater surface-volume-ratio, thus are more rapidly biodegradable and might be more biological available for smaller non-target organisms in the food chain. For limiting exposure of harvest by-products to protected aquatic ecosystems in respect to the precautionary principle and minimising potential adverse impacts of Bt-maize on aquatic organisms, buffer zones along protected rivers and lakes seem to be potentially effective measures. The investigation will be carried on in 2010 by further field experiments and modelling of the input.

References

Bøhn T., Primicerio R., Hessen D.O., Traavik, T. (2008) Reduced fitness of Daphnia magna fed a Bt-transgenic maize variety. Arch. Environ. Contam. Toxicol. 55(4): 584–592.

Grimmert S., Harders H., Vögel R., Peil J. (2009) Biogasanlagen und Maisanbau in Brandenburg, Entwicklung von 2005–2009. Gülzower Fachgespräche 32: 442–443.

Jensen P.D., Dively G.P., Swan C.M., Lamp W.O. (2010) Exposure and nontarget effects of transgenic Bt corn debris in streams. Envir. Entomol. 39(2): 707–714.

Rosi-Marshall E.J., Tank J.L., Royer T.V., Whiles M.R., Evans-White M., Chambers C., Griffiths N.A., Pokelsek, J., Stephen, M.L. (2007) Toxins in transgenic crop byproducts may affect headwater stream ecosystems. PNAS 104(41): 16204–16208.

Networks and environmental observation programs as a tool for general surveillance – first experience and future requirements

Wiebke Zughart
(Federal Agency for Nature Conservation, Bonn, Germany. –
Wiebke.Zueghart@BfN.de)

Abstract

Consent holders make use of the opportunity to involve existing observation programs or networks in the general surveillance of GMOs. Three core strategies are currently established: the participation of European trade organisations, contributions of designated experts and the assessment of data gathered by environmental observation programs operated by third parties. In this contribution, needs for improvement are identified based on the analysis of monitoring plans and reports. Reported results and conclusions drawn by the consent holder are often neither traceable nor assessable because of the lack of explanations of monitoring objectives, methods and data analysis. To assure a reliable general surveillance of GMOs, science-based criteria for the selection of appropriate programs and networks as well as a data quality management are essential and must be developed. Agreements concerning the availability of data have to be settled before consent for placing GMOs on the market can be given.

Introduction

It is recommended by European legal provisions to use existing monitoring schemes for the general surveillance of GMOs. The guidance notes to Annex VII of Directive 2001/18/EC state that

"GS could, where compatible, make use of established routine surveillance practices such as monitoring of agricultural crops, plant protection, veterinary and medical products as well as ecological monitoring, environmental observation and nature conservation programs" (EC 2002). The guidance notes continue that

"If established routine surveillance practice is used in the general surveillance, this practice should be described as well as the changes in the practice needed to fulfill a relevant general surveillance" (EC 2002).

All currently established general surveillance plans make use of this opportunity (EFSA 2009). Although the approaches applied are quite different from each other, they all reveal strong deficiencies and require fundamental improvement.

Strategies

At present, three core strategies are implemented: participation of European trade associations, contribution of designated experts and the assessment of data gathered by environmental observation programs operated by third parties.

European trade organisations
European trade associations like COCERAL (importers/traders), UNISTOCK (silo operators) or FEDIOL (processors) are involved in the monitoring of crops approved for import and processing, feed or food uses (EFSA 2009). The idea is that the associations inform and remind their member organisations and companies annually
- to monitor for adverse effects,
- to inform their own member companies of this requirement and
- to report any findings to the European trade association.

The European trade association will report directly or via EuropaBio to the party who is holder of the consent to release the GMO (consent holder).

It is remarkable that neither the monitoring plans nor the reports give any information concerning the monitoring procedure. It remains unclear who participated in the monitoring and what kind of response is generated. No details on monitoring objectives, methods, locations, frequencies or expertise of participants are given (EFSA 2009). Hence, the monitoring conducted by European trade associations and the reported results are neither traceable nor assessable.

Designated experts
In the case of imported genetically modified carnation stems the general surveillance of potential environmental effects is carried out by experts (EFSA 2009). Three breeders and six botanists who are concerned with *Dianthus* biology were asked to alert the consent holder to any unusual hybrids that are found during their routine surveys. In addition, the consent holder asked herbaria, national botanical survey networks, plant protection services and botanical gardens in Europe to be alerted in case of dispersal of GM carnation or the occurrence of hybrids. To benefit from the knowledge of designated experts is a step forward. However, the participation of the above mentioned experts and institutions is voluntary and there are no binding agreements. Therefore, no systematic observations are conducted and any findings and reports will occur by chance.

Environmental observation programs
A "German network monitoring" was implemented in Germany in 2008 during the cultivation of MON810. Its main strategy is to review the reports published annually by selected environmental observation programs (BVL 2008) in order to find out if any adverse effects of MON810 cultivation can be identified. Whenever adverse effects are recognized, the consent holder will contact the corresponding observation programs and ask for the relevant primary data to analyse them.

The monitoring report (Monsanto 2009) delivered in March 2009 showed that this strategy has failed and needs fundamental improvement. Only some of the selected environmental observation programs publish their data in publicly available reports. Even if results are available, they do not necessarily provide relevant information. The observation programs were established for other purposes than monitoring environmental effects of GMOs. Thus, scope and parameters, time, frequency and scale of data collection as well as the methods for sampling and analysis do not fit into the task of GMO monitoring.

Another problem of this strategy is that program coordinators or responsible persons are not contacted beforehand about an agreement on the delivery of data. Collectors of data, who are often volunteers, may show reluctance to provide their data to consent holders, as some did in 2008 (Agrarheute 2009a,b).

Conclusions

In principle, existing networks, services and environmental observation programs could make a valuable contribution to general surveillance (Züghart et al. 2008). However, precise and science-based criteria are needed to select appropriate networks and observation programs. Options for adaptation or enlargement of programs are to be considered. If existing networks and programs are not suitable, additional monitoring tools or surveys have to be implemented.

In order to ensure a suitable and sound data base, quality management is crucial. Thus, clearness concerning observation objects, monitoring design, experts involved, information flow and data analysis is essential. Agreements concerning access to data or results should be settled before the authorisation for placing a GMO on the market is granted.

If significant effects on human health or on the environment are reported, in-depth studies should be carried out to determine the causes (EFSA 2006). However, it is still not defined in which case further studies are indicated, how such studies should be designed and who will be responsible for their conduct. Therefore, a process that allows clear and fast responses to findings from environmental observation programs or networks must be developed.

References

Agrarheute (2009a) Monitoring ist nicht geeignet, um die Auswirkungen von MON810 zu bestimmen. http://www.agrarheute.com/index.php?redid=291774.

Agrarheute (2009b) Monsanto nutzte Daten ohne Absprache. http://www.agrarheute.com/index.php?redid=292685.

BVL (2008) Monitoringplan GV Mais MON810, Part 1: page 114–147, Part 2: page 148–160. http://www.bvl.bund.de/cln_027/nn_491652/DE/08__PresseInfothek/00__doks__downloads/Monitoringplan.html.

EC (2002) Council Decision of 3 October 2002 establishing guidance notes supplementing Annex VII to Directive 2001/18/EC of the European Parliament and of the Council on the deliberate release into the environment of genetically modified organisms and repealing Council Directive 90/220/EEC. Official Journal of the European Communities, L 280/27, 2002/811/EC.

EFSA (2006) Guidance document of the Scientific Panel on Genetically Modified Organisms for the risk assessment of genetically modified plants and derived food and feed. The EFSA Journal 99: 1–100.

EFSA (2009) Register of questions. http://registerofquestions.efsa.europa.eu

Monsanto Company (2009) 2008 German Network Monitoring. https://yieldgard.eu/en-us/YieldGardLibraryGrower/2008%20Yieldgard%20German%20Network%20Monitoring%20Report.pdf.

Züghart W., Benzler A., Berhorn F., Sukopp U., Graef F. (2008) Determining indicators, methods and sites for monitoring potential adverse effects of genetically modified plants to the environment: the legal and conceptional framework for implementation. Euphytica 164 (3): 845–852.

Breckling, B. & Verhoeven, R. (2010) Implications of GM-Crop Cultivation at Large Spatial Scales.
Theorie in der Ökologie 16. Frankfurt, Peter Lang.

Genetically modified crop consumption at large scale: Possible negative health impacts due to holes in assessment. Overview of the safety studies of GMOs performed on mammals[1]

Gilles-Eric Séralini[a,b], Joël Spiroux de Vendomois[b], Dominique Cellier[b,c], Robin Mesnage[a,b] & Emilie Clair[a,b]
([a]University of Caen, Risk Pole CNRS, Caen cedex; [b]CRIIGEN, Paris; [c]LITIS, University of Rouen; France. – criigen@unicaen.fr)

Background, aim and scope

Recently, a debate on international regulation is ongoing on the capacity to predict and avoid adverse effects on health and environment of new products and novel food/feed (GMOs, chemicals, pesticides, nanoparticles …). The health risks assessment cannot avoid the study of blood analyses of mammals eating these products in subchronic or chronic tests. Mammalian feeding trials have thus been usually performed for regulatory purposes, in order to obtain authorizations or commercialization for GM plant derived foods or feed. They may have been published in the scientific literature afterwards. We have obtained, following Court actions or official requests, the raw data of several safety, 28 day or 90 day long, in vivo tests on rats for GMOs (Séralini et al. 2007; Spiroux de Vendomois et al. 2009). We have thoroughly reviewed these tests from both a biological and biostatistical point of view. We focus here on the results of available 90-day feeding trials (or more) with commercialized GMOs, in the light of modern scientific knowledge.

Overview of the safety studies performed on mammals

Firstly we have focused our study on commercialized GMOs which have been cultivated in significant amounts throughout the world since 1994. These often analyze the biochemical blood and urine parameters of mammals eating GMOs, together with numerous organ weights and histopathology. We observe and emphasize that all the events correspond to soybean and maize which constitute 83 % of the commercialized GMOs, whilst the remainder are canola or cotton (Clive 2009). Then, some tests presented here show controversial results which can be discussed, or statistically significant results considered as not biologically significant by regulatory authorities, raising the question of the statistical and biological interpretation of results.

---

1 Extended abstract: A full paper is submitted to UWSF – Zeitschrift für Umweltchemie und Ökotoxikologie, Series: Implications of GMO-cultivation and monitoring. Springer-Verlag.

First of all, the data indicating no biological significance of statistical effects in comparison to controls have been published mostly by companies from 2004, at least 10 years after commercialization in the world of these GMOs. This is a matter of grave concern. Moreover, only three events were tested more than 90d long or on more than one generation, and not by industry; even if a 90d period is considered as insufficient to evaluate chronic toxicity (Séralini et al. 2009; EFSA 2008). All these commercialized cultivated GMOs have been modified to contain pesticides, either by herbicide tolerance or insecticide production, or both, and thus could be considered as "pesticide plants" (Seralini 2004). These GMOs encode only for these two traits in spite of the advertising for numerous other characters possibly existing. Usually, pesticides are tested over a period of 2 years on a mammal to measure quite often side effects. Additionally, unintended effects of the genetic modification itself cannot be excluded, as direct or indirect consequences of insertional mutagenesis creating possibly metabolic effects (Rosati et al. 2008).

Some GMOs affect the body weight increase according to the authors (RR CP4 EPSPS and MON863) at least in one sex (Séralini et al. 2007; Zhu et al. 2004), a parameter considered as a very good predictor of side effects in other organs. Several convergent factors appear to indicate liver and kidney problems as endpoints of GMO diet effects in these experiments (Séralini et al. 2007; Spiroux de Vendomois et al. 2009; Vecchio et al. 2004; Kilic and Akay 2008). This is confirmed by our meta-analysis of all in vivo studies published on this topic (Table 1). On a total of around 9 % of all parameters disrupted, 2 times more than that could be obtained by chance only, surprisingly 42 % of those were concentrated on male kidneys for all commercialized GMOs, even if only around 25 % of the parameters measured concerned directly the kidney level. Even if our own counter-expertise is removed from the calculation, showing numerous kidney dysfunctions (Séralini et al. 2007), around 32 % of disturbances are still noticed in kidneys. However, other organs maybe reached such as heart and spleen, or blood cells (Spiroux de Vendomois et al. 2009).

Tab. 1: Meta-analysis of statistical differences of feeding trials with commercialized soybean and maize GMOs given to rats (Séralini et al. 2010). Parameters are classified per tissues according to Séralini et al. (2007). Statistical differences are reported according to the statistics of the authors. All these data revealed that the kidney is particularly reached, concentrating 42 % of all parameters disrupted in males.

Parameters in GMOs in vivo studies of toxicity	Measured by organ (%) / Total (721-719)		Disturbed in each organ (%) / Total disrupted parameters (~ 10 %)	
	Females	Males	Females	Males
Liver	23.4	23.5	26	24.3
Kidney	23.9	23.9	27.5	41.4
Bone Marrow	30.8	30.9	34.8	27.1
Total for 3 tissues	88.1	88.3	88.3	93.4

Conclusion

We can conclude from regulatory tests performed today that it is unacceptable to submit 500 million Europeans and several billions of consumers worldwide to these new pesticide-GM derived foods or feed, and this without more controls than if any only 3 month long toxicological tests, and this with only one mammalian species, especially given the evidence of worrying problems.

References

Clive J. (2009) Global Status of Commercialized Biotech/GM Crops: 2009. ISAAA Brief 41.
EFSA (2008) Safety and nutritional assessment of GM plants and derived food and feed: the role of animal feeding trials. Food Chem Toxicol 46: 2–70.
Hammond B., Lemen J., Dudek R., Ward D., Jiang C., Nemeth M., et al. (2006) Results of a 90-day safety assurance study with rats fed grain from corn rootworm-protected corn. Food Chem Toxicol 44: 147–160.
Kilic A., Akay M.T. (2008) A three generation study with genetically modified Bt corn in rats: Biochemical and histopathological investigation. Food Chem Toxicol 46: 1164–1170.
Rosati A., Bogani P., Santarlasci A., Buiatti M. (2008) Characterisation of 3' transgene insertion site and derived mRNAs in MON810 YieldGard maize. Plant Mol Biol 67: 271–281.
Séralini G.E. (2004) Ces OGM qui changent le monde, Flammarion.
Séralini G.E., Cellier D., Spiroux de Vendomois J. (2007) New analysis of a rat feeding study with a genetically modified maize reveals signs of hepatorenal toxicity. Arch Environ Contam Toxicol 52: 596–602.
Séralini G.E., Spiroux de Vendomois J., Cellier D., Sultan C., Buiatti M., Gallagher L., et al. (2009) How subchronic and chronic health effects can be neglected for GMOs, pesticides or chemicals. Int J Biol Sci 5: 438–443.
Séralini G.E., Mesnage R., Clair E., Spiroux de Vendômois J., Cellier D. (2010) Genetically modified crops consumption at large scale: possible negative health impacts due to holes in assessment. Umweltwiss Schadst Forsch. Submitted.
Spiroux de Vendomois J., Roullier F., Cellier D., Seralini G.E. (2009) A comparison of the effects of three GM corn varieties on mammalian health. Int J Biol Sci 5: 706–726.
Vecchio L., Cisterna B., Malatesta M., Martin T.E., Biggiogera M. (2004) Ultrastructural analysis of testes from mice fed on genetically modified soybean. Eur J Histochem 48: 448–454.
Zhu Y., Li D., Wang F., Yin J., Jin H. (2004) Nutritional assessment and fate of DNA of soybean meal from roundup ready or conventional soybeans using rats. Arch Anim Nutr 58: 295–310.

Breckling, B. & Verhoeven, R. (2010) Implications of GM-Crop Cultivation at Large Spatial Scales. Theorie in der Ökologie 16. Frankfurt, Peter Lang.

Roundup® in genetically modified plants: Regulation and toxicity in mammals[1]

Robin Mesnage, Emilie Clair & Gilles-Eric Séralini
(University of Caen, Institute of Biology, CRIIGEN and Risk Pole CNRS, Caen Cedex France. – criigen@unicaen.fr)

Context

Among the 134 million hectares of genetically modified plants growing worldwide in 2009, more than 99.9 % are described as pesticide plants (Clive 2009). Around 80 % are tolerant to Roundup, a glyphosate based herbicide. Its use on GMOs is thus amplified, and this phenomenon shed a new light on the problem of herbicide residues in plants. This is because these GM plants have been modified so that they can contain high levels of Roundup. They are modified to behave normally after several treatments with this herbicide, which were not allowed at such levels on regular plants before. The latest generation, like Smartstax crops, even cumulate a tolerance up to 2 herbicides and a production of 6 insecticides. By this widespread use and the known potential hazards of pesticides, their residues are a major concern for health and the environment. Moreover the new metabolism that they could trigger in GMOs remains to be studied. A debate on international standards is ongoing on their capacity to predict and avoid adverse effects of the herbicide residues at environmental or nutritional exposures, particularly in GMOs.

As far as Roundup is concerned, the formulations of which are mixtures of only one proposed active ingredient (glyphosate) with various adjuvants, up to 400 ppm of residues are authorized in some Genetically Modified food and feed (EPA 2008). It is also recognized by regulatory agencies that these residues are found in meat and products generated from livestock fed with glyphosate tolerant soya or maize (EFSA 2009).

Review on Roundup toxicity studies

Surprisingly, more and more studies have revealed unexpected effects of Roundup, including carcinogenic and endocrine disrupting effects. This is at lower doses than those authorized for residues found in Genetically Modified Organisms (GMOs). For example, Roundup altered the spermatogenesis of rats exposed in utero to 50 ppm per day (Dallegrave et al. 2007). Even a tumour promoting potential is observed on mice

---

[1] Extended abstract: A full paper is submitted to UWSF – Zeitschrift für Umweltchemie und Ökotoxikologie, Series: Implications of GMO-cultivation and monitoring. Springer-Verlag.

exposed to 25 ppm per day (George et al. 2010). Alterations of rat testicular morphology and testosterone levels occur at doses of 5 ppm per day (Romano et al. 2009). In our laboratory we have observed endocrine disruption on human cell lines; it was a disruption of aromatase, of the androgen and estrogen receptors in 24 hours, starting from 0.5 ppm Roundup. This corresponds to glyphosate concentrations 2000 times less than the authorized levels in GMOs (Gasnier et al. 2009). Furthermore, we have shown that Roundup inhibited cellular respiration, and that it also caused membrane damages. Last but not least, Roundup showed genotoxic effects, as well as it induced apoptosis and necrosis in human cells (Benachour & Séralini 2009). Most of these effects are amplified with time. This is preoccupying, and it does highlight the limits of the Acceptable Daily Intake concept for long term exposures.

Debate on health risks

In all these studies, toxic effects were not detected with the so-called active ingredient glyphosate alone at these doses; they were more related to the formulations of the herbicide and its adjuvants. These remain confidential and their residues are not measured. Out of the 20 tests required (or conditionally required) to register a pesticide in the United States, only 7 short-term acute toxicity tests use the whole formulation; the others are done using the sole active ingredient (Cox & Surgan 2006). The problem of pesticide registration is indeed very old, and it is only the active ingredient that is tested in chronic mammalian toxicity tests (generally for 2 years on rats). Moreover there is generally only one 2-year test worldwide on a mammal per pesticide, performed by the company commercializing this pesticide. Adjuvants are often considered to be inert in the assessment process. This is a major issue. Such a simplistic approach of pesticides hazards bypasses the potential effects of adjuvants and their mixtures with the active ingredient on chronic risks. This issue is even more crucial with GMOs which are designed to tolerate the formulations that enter the edible plant cells.

Nevertheless, it is well known that adjuvants are mixed with the active ingredient in order to increase the efficiency of formulations. In medicine, adjuvants are also used to increase the molecule absorptions, or the effectiveness of vaccines. In chemical products such as pesticides, they are used to increase targeted toxicity (for example penetration in leaves or insects), but they do have an effect also on non specific targets too. Some known adjuvants of Roundup such as polyethoxylated tallowamine (or POEA) showed more toxic effects than glyphosate in various models, and even more than Roundup in some cases on aquatic life for example (Tsui & Chu 2003; Marc et al. 2005) or on human cells (Benachour & Seralini 2009).

By only considering the active ingredient, regulatory thresholds seem to guarantee the safety of residues, however we conclude that it is not the case with the whole formulations, in particular those specific to GMOs. In conclusion, confidentiality on the composition of formulations must be lifted, as announced recently by the U.S. Environmental Protection Agency following our work (EPA 2009). People consuming GMOs are thus exposed to residues of many formulations which are themselves mixtures of different chemicals. The long term combined effects have never been evaluated, not even in laboratory animals. We suggest that regulatory agencies change their paradigms and integrate modern knowledge, in order to guarantee the safety of pesticides residues, in particular when associated with genetically modified plants.

References

Benachour N., Seralini G.E. (2009) Glyphosate formulations induce apoptosis and necrosis in human umbilical, embryonic, and placental cells. Chem Res Toxicol 22: 97–105.
Clive J. (2009) Global Status of Commercialized Biotech/GM Crops: 2009. ISAAA Brief 41.
Cox C., Surgan M. (2006) Unidentified inert ingredients in pesticides: implications for human and environmental health. Environ Health Perspect 114: 1803–1806.
Dallegrave E., Mantese F.D., Oliveira R.T., Andrade A.J., Dalsenter P.R., Langeloh A. (2007) Pre- and postnatal toxicity of the commercial glyphosate formulation in Wistar rats. Arch Toxicol 81: 665–673.
EFSA (2009) Modification of the residue definition of glyphosate in genetically modified maize grain and soybeans, and in products of animal origin on request from the European Commission. EFSA Journal 7: 42.
EPA (2008) Federal Register / Rules and Regulation. 73588–73592.
EPA (2009) Glyphosate final work plan (FWP) Registration review case No. 0178.
Gasnier C., Dumont C., Benachour N., Clair E., Chagnon M.C., Seralini G.E. (2009) Glyphosate-based herbicides are toxic and endocrine disruptors in human cell lines. Toxicology 262: 184–191.
George J., Prasad S., Mahmood Z., Shukla Y. (2010) Studies on glyphosate-induced carcinogenicity in mouse skin: A proteomic approach. J Proteomics. 73: 951–965.
Marc J., Le Breton M., Cormier P., Morales J., Belle R., Mulner-Lorillon O. (2005) A glyphosate-based pesticide impinges on transcription. Toxicol Appl Pharmacol 203: 1–8.
Romano R.M., Romano M.A., Bernardi M.M., Furtado P.V., Oliveira C.A. (2009) Prepubertal exposure to commercial formulation of the herbicide glyphosate alters testosterone levels and testicular morphology. Arch Toxicol. DOI: 10.1007/s00204-009-0494-z.
Tsui M.T., Chu L.M. (2003) Aquatic toxicity of glyphosate-based formulations: comparison between different organisms and the effects of environmental factors. Chemosphere 52: 1189–1197.

Seedling emergence of oilseed rape (*B. napus* L.) and wild relatives on ruderal soils

Jana Seeger, Broder Breckling & Juliane Filser
(University of Bremen, UFT Centre of Environmental Research and Sustainable Technology, Bremen, Germany. – jseeger@uni-bremen.de)

Abstract

We assessed seedling emergence of *Brassica napus* and *B. rapa* in comparison with their weedy relatives *B. nigra* and *Raphanus raphanistrum*. Plant performance was tested on four ruderal substrates on a former rubble dump site in Bremen, Germany. The effect of soil type on seedling emergence was small and varied between plant species. However, the cultivated plants performed substantially better than the weedy plants on each substrate. This raises concern for the uncontrolled spread of transgenes from GM plants via ruderal populations.

Introduction

The ability of *Brassica napus* to develop feral populations, not only on arable land but also on ruderal sites (Menzel 2006), is an important aspect in risk assessment of GM oilseed rape. Our study aimed at determining the magnitude of seedling emergence on ruderal soils and comparing this with the performance of weedy relatives. We chose closely related crucifer species, all of which *B. napus* can produce fertile F1 hybrids with (Scheffler & Dale 1994). *B. rapa* occurs in weedy and cultivated varieties and is most successful in hybridisation with *B. napus* (Scheffler & Dale 1994). *Raphanus raphanistrum* and, particularly in temperate North America, *Brassica nigra*, are both widespread and problematic weeds typically growing in disturbed habitats (Callihan et al. 2000; Holm et al. 1977). In such weeds, selection is likely to favour tolerance for a high amplitude of environmental conditions (Holzner 1982). Thus, we expected them to perform better on ruderal soils than the cultivated plants.

Methods

A fully randomised two-factorial split-plot design with 20 replicates was installed on a former dump site for building rubble in Bremen, Germany. 20 blocks with four plastic containers each (Ø 53 cm, open bottom) were set up and protected from slugs and small mammals by fences. The containers were dug in and filled to a depth of 34 cm with one of four substrates: sand, mixed soil, humous soil and shallow soil (Table 1). Each plant species was assigned to one pie-shaped fourth of each container, into which 90 seeds of the plant were sown in late October 2007. A sowing depth of 1 cm was chosen to minimize seed predation and to enhance seedling emergence of *R. raphanistrum* (Cheam 1986). Seeds originated from cultivated varieties (*B. napus* and *B. rapa*) or from ruderal populations (wild relatives). Total seedling emergence [%] was calculated from the maximum number of individuals found over three censuses from November 2007 to May 2008.

Tab. 1: Properties of the four ruderal substrates. WHC_{max} = maximum water holding capacity (% dm); SOM = soil organic matter content (% dm); dm = dry weight.

substrate	description	SOM	WHC_{max}	pH
sand	from a ruderal site	0.2	20.6	4.1
mixed	1:1 sand and humous soil	1.8	25.2	6.6
humous soil	from the dump site	3.8	36.6	6.7
shallow humous soil	9 cm layer of humous soil over building rubble	3.8	36.6	6.7

Results and discussion

Differences in seedling emergence between the four plant species were significant for all plant comparisons (Table 2, Tukey post-hoc tests, $p < 0.05$). Unlike excepted, seedling emergence was substantially higher for the cultivated species *B. napus* (82 %, Figure 1) and *B. rapa* (66 %) than for their wild relatives *B. nigra* (28 %) and *R. raphanistrum* (15 %). This was also true for the percentage of fruiting individuals and the reproductive output (manuscript in preparation). The poor performance of the wild relatives might partly be explained by a) a potentially high level of dormancy in *R. raphanistrum* (Roberts & Boddrell 1983), leading to low seedling emergence and b) the late sowing date disfavouring the comparably small-seeded *B. nigra* through higher seedling mortality in the winter.

Tab. 2: Results of a two-way split-plot ANOVA to analyse the effect of plant species (between-blocks factor) and substrate (within-blocks factor) on seedling emergence [%] (arcsine square root transformed prior to analysis).

Source of variation	df	Mean Square	F	p
Block	19	29.73	1.21	
Substrate	3	339.85	13.80	<0.001
error (main plot)	57	24.62	0.88	
plant sp.	3	30008.66	1072.70	<0.001
Substrate*plant sp.	9	246.99	8.83	<0.001
error (sub-plot)	228	27.97		

Substrate comparisons revealed significant effects for all plant species (Figure 1, One-way blocked ANOVAs on arcsine square root transformed data. *B. napus*: $p = 0.025$,

$F_{3,76} = 3.37$; *B. rapa*: $p < 0.001$, $F_{3,76} = 16.15$; *B. nigra*: $p < 0.001$, $F_{3,76} = 13.17$; *R. raphanistrum*: $p = 0.042$, $F_{3,76} = 2.92$). Significant negative effects of low-quality soil (sand/shallow) on seedling emergence were found for *B. rapa*, but not for *B. napus* (Figure 1). Although substrate effects differed between species, seedling emergence was never lower for the cultivated plants than for the wild relatives. This suggests that *B. napus* may cope with low-quality ruderal soils at least as well as weedy relatives in terms of seedling emergence. Further analysis of the fruiting plants confirmed this (Seeger et al., in preparation), showing that *B. napus* was able to produce seeds on all soils. Its total seed production was overall significantly higher than that of the wild relatives. The demonstrated potential weediness of *B. napus* increases the risk of uncontrolled spread of transgenes through feral populations.

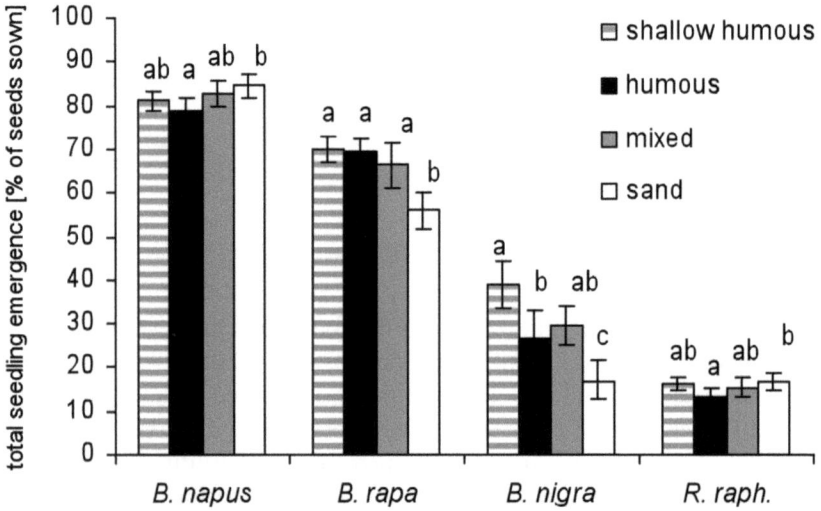

Fig. 1: Seedling emergence [%] per plant species on the different substrates (back-transformed means ± 95 % confidence intervals). Different letters indicate significant differences between substrates (p < 0.05, Tukey post-hoc tests).

References

Callihan B., Brennan J., Miller T., Brown J., Moore M. (2009) Mustards in Mustards: Guide to Identification of Canola, Mustard, Rapeseed and Related Weeds. Moscow ID, University of Idaho.

Cheam A.H. (1986) Seed production and seed dormancy in wild radish (*Raphanus raphanistrum* L.) and some possibilities for improving control. Weed Research 26: 405–413.

Holm L.G., Plucknett D.L., Pancho J.V., Herberger J.P. (1977) The World's Worst Weeds – Distribution and Biology. Honolulu, University Press of Hawaii.

Holzner W. & Numata M. (1982) Biology and ecology of weeds. The Hague, Dr. W Junk Publ.

Menzel G. (2006) Verbreitungsdynamik und Auskreuzungspotenzial von Brassica napus L. (Raps) im Großraum Bremen. GCA-Verlag, Waabs, Dissertation University of Bremen.

Roberts H.A., Boddrell J.E. (1983) Seed survival and periodicity of seedling emergence in eight species of Cruciferae. Annals of Applied Biology 103: 301–304.

Scheffler J.A., Dale P.J. (1994) Opportunities for gene transfer from transgenic oilseed rape (Brassica napus) to related species. Transgenic Research 3: 263–278.

Breckling, B. & Verhoeven, R. (2010) Implications of GM-Crop Cultivation at Large Spatial Scales.
Theorie in der Ökologie 16. Frankfurt, Peter Lang.

Setup, efforts and significance of a GMO monitoring program – An Austrian case study[1]

Kathrin Pascher, Dietmar Moser, Stefan Dullinger, Leopold Sachslehner, Patrick Gros, Norbert Sauberer, Andreas Traxler, Georg Grabherr & Thomas Frank
(Department of Conservation Biology, Vegetation Ecology and Landscape Ecology (CVL), Vienna, Austria. – kathrin.pascher@univie.ac.at)

Abstract

A monitoring procedure is part of the precautionary principle of genetically modified crop cultivation. It is mandatory according to the Directive 2001/18/EC. "Case-specific monitoring serves to confirm that scientifically sound assumptions, in the environmental risk assessment, regarding potential adverse effects arising from a GMO (genetically modified organism) and its use are correct. General surveillance is largely based on routine observation ("look – see" approach) and should be used to identify the occurrence of unforeseen adverse effects of the GMO or its use for human health and the environment that were not predicted in the risk assessment (Council Decision 2002/811/EC)".

The British Farm Scale Evaluations demonstrated that reductions in species numbers and diversity of agro-ecosystems may be among these unforeseen adverse effects of genetically modified plants (GMPs) (e.g. Bohan et al. 2005). Hence, biodiversity is an important object of a general surveillance program on ecological impacts of GMP cropping.

A GMP monitoring program in the sense of a general surveillance should comply with (1) providing baseline data for detecting unintended long-term effects on the abundance and diversity of plants and animals as well as changes in habitat structures, (2) describing general agricultural trends on biodiversity, (3) estimating the magnitudes of any difference in biodiversity, (4) detecting and assigning specific GMP effects, (5) allowing an interpretation of changes in important ecological processes, (6) choosing appropriate indicators and parameters to sense unexpected impacts, (7) providing test areas representative for the ranges of soil types, climatic conditions and management regimes of a country, (8) performing a first risk assessment of special GMPs based on the occurrence and abundance of related species and (9) providing a framework into which additional indicators can be integrated for an extended survey (Bühler 2006; Firbank et al. 2003; Pascher et al. 2009).

---

1 Extended abstract: A full paper is submitted to UWSF – Zeitschrift für Umweltchemie und Ökotoxikologie, Series: Implications of GMO-cultivation and monitoring. Springer-Verlag.

An Austrian GMP monitoring program BINATS (Biodiversity-Nature-Safety) was developed and implemented on 100 representative test areas all over Austria (Pascher et al. 2008; 2009). Vascular plants, grasshoppers, butterflies and habitat structures were chosen as indicators. Data on the diversity of these taxa were collected to provide essential baseline information. This first monitoring cycle revealed insights into both, the significance and the limits of such a monitoring program and allowed for a realistic calculation of the associated costs.

References

Bohan D.A., Boffey C.W.H, Brooks D.R., Clark S.J. et al. (2005) Effects on weed and invertebrate abundance and diversity of herbicide management in genetically modified herbicide-tolerant winter-sown oilseed rape. Proceedings of the Royal Society B 272: 463–474.

Bühler C. (2006) Biodiversity Monitoring in Switzerland: What can we learn for general surveillance on GM crops? Journal für Verbraucherschutz und Lebensmittelsicherheit 1, Supplement 1: 37–41.

Council Decision 2002/811/EC of 3 October (2002) Establishing guidance notes supplementing Annex VII to Directive 2001/18/ EC of the European Parliament and of the Council on the deliberate release into the environment of genetically modified organisms and repealing Council Directive 90/220/EEC.

Firbank L.G., Heard M.S., Woiwod I.P., Hawes C. et al. (2003) An introduction to the Farm-Scale Evaluations of genetically modified herbicide-tolerant crops. Journal of Applied Ecology 40: 2–16.

Pascher K., Moser D., Dullinger S., Sachslehner L. et al. (2008) Monitoring design to evaluate biodiversity in Austrian agricultural regions. In: Breckling B., Reuter H. & Verhoeven R. (eds.): Implications of GM-Crop Cultivation at Large Spatial Scales. Theorie in der Ökologie 14. Frankfurt, Peter Lang: 146–150. www.gmls.eu.

Pascher K., Moser D., Dullinger S., Sachslehner L. et al. (2009) Establishment of an Austrian monitoring design of genetically modified plants. Proceedings of the Fourth International Conference on Coexistence between Genetically Modified (GM) and non-GM based Agricultural Supply Chains. GMCC09, 10–12th November 2009, Melbourne, Australia: 11 pp. www.gmcc-09.com.

Breckling, B. & Verhoeven, R. (2010) Implications of GM-Crop Cultivation at Large Spatial Scales. Theorie in der Ökologie 16. Frankfurt, Peter Lang.

Prioritizing GMO monitoring sites in dynamic cultivation systems and their environment – a conceptual on-farm approach

Claudia Bethwell, Frieder Graef, Ulrich Stachow, Hans-Jürgen Müller & Frank Eulenstein
(Leibniz Centre for Agricultural Landscape Research (ZALF), Dept. for Land Use Systems, Müncheberg, Germany – Claudia.Bethwell@zalf.de)

Introduction

The release of genetically modified organisms (GMOs) into EU-environments must be accompanied by a monitoring to detect potential adverse effects (Directive 2001/18/EC). The monitoring has to take place in exposed areas, i.e. cultivated fields and their environment (Züghart et al. 2008).

One question in GMO monitoring is where to locate monitoring sites on-farm, given that on the one hand the crop cultivation systems and their environments are diverse and dynamic and on the other that monitoring should be affordable. Therefore knowledge of dynamics and spatial patterns in GM crop cultivation is required.

Since in Germany currently there is no large-scale GM crop cultivation we used the present state of maize cultivation in Brandenburg as a scenario for the potential spatial distribution of Bt-maize cultivation. We selected four typical farms in Brandenburg State with maize cultivation and used their real cultivation data as a scenario for a potential GM Bt-maize cultivation. We also included Corine Landcover data (Bossard et al. 2000) to achieve information from surrounding biotopes. Regarding the spatio-temporal field use variability of the four farms we developed a sequential scheme for determining preferable monitoring sites.

The objective of this methodical paper is to present the general approach, including (a) the GIS-algorithm for preparing the data and analysing the spatio-temporal patterns and dynamic of maize cultivation, (b) the assessment scheme for determining the potential exposure of Bt-maize on farm fields and surrounding biotopes as a prerequisite to (c) develop an on-farm monitoring scheme.

Study area and data

The study area is located in the administrative district Märkisch-Oderland in the eastern part of Brandenburg (Figure 1). It includes the Oder valley and several moraine plateaus. The region is characterized by about 118,000 ha of arable land with 9,600 ha of maize cultivation (8 % of arable land) (Statistische Ämter des Bundes und der Länder, 2009). Before the temporary ban of Bt-Maize MON810 in Germany in 2009 Märkisch-Oderland has been the centre of Bt-maize cultivation with 550 ha of Bt-maize cultivation in 2007 (0.5 % of arable land) (MLUV 2008; BVL 2010).

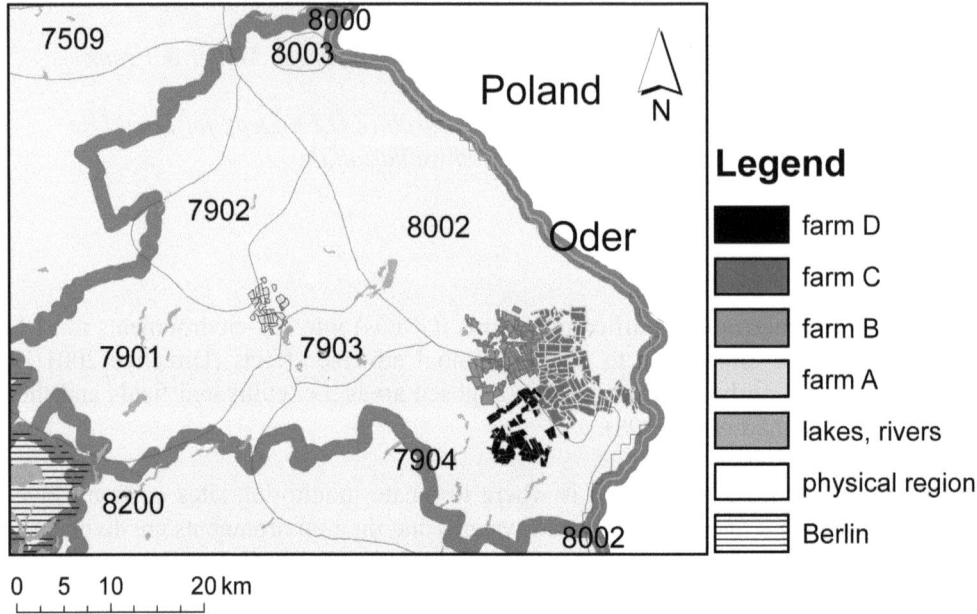

Fig. 1: The Brandenburg administrative district "Märkisch-Oderland", the selected farms, and physical regions according to Meynen & Schmidthüsen (1962) for instance the Oder valley, 8002 (Oderbruch) and moraine plateaus, 7901 (Barnimplatte) and 7904 (Lebusplatte).

We used crop cultivation data of four farms in Märkisch-Oderland. This included a database with field specific crop cultivation data from 2002 to 2007 and a digital map of farms and field locations. The four farms are partly situated in the Oder valley and in the moraine plateaus. The average field size is between 28 ha (farm A) and 66 ha (farm C).

Development of a sequential scheme for determining preferable monitoring sites

We focused on present maize cultivation as a scenario for potential Bt-maize cultivation including refugial areas and buffer planting according to best agricultural practice and experience with cultivation of MON810 in Brandenburg (LVLF Brandenburg 2005). We discerned three land use classes on and around the four farms: Maize fields, other cropped fields, and the surrounding biotopes. For the exposure of these land use classes to GMOs we considered two categories: Maize fields and areas neighbouring maize fields.

We preprocessed the data and developed a GIS-algorithm using ArcGIS to analyse the spatio-temporal maize cropping patterns and dynamic (Figure 2). We applied different exposure distances using an outer buffer. The spatio-temporal concentration we achieved by intersecting land use categories and buffer areas for every year and by intersection of the single years.

For determining the potential Bt-maize exposure in more detail we categorised various different intensities: a) distant non-maize fields and biotopes, b) areas neighbouring maize fields such as surrounding biotopes or non-maize fields, and c) maize fields. We then calculated the differing temporal exposure concentrations with repeated Bt-maize cultivation on the same field.

Land use and maize cultivation

A first data analysis provides an overview of land use and maize cultivation intensity on the farms and of their surrounding biotopes. The land use on the farm areas and neighbouring biotopes was derived from Corine Landcover 2000 (CLC 2000) and calculated with 50 m, 200 m and 1000 m buffer areas (Table 1). Apart from the dominant agricultural areas there are minor proportions of areas with permanent crops and grassland (class 2.2 and 2.3), urban areas (class 1), forests and semi-natural areas (class 3), wetlands (class 4) and water bodies (class 5). The area of non-agricultural land or biotopes increases with more distance from farm land.

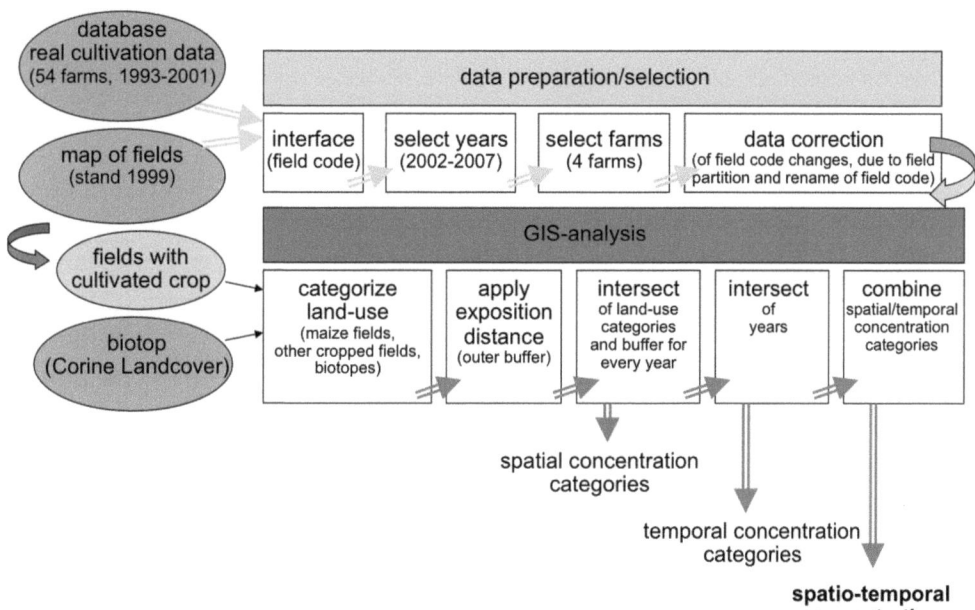

Fig. 2: Data processing and GIS-algorithm.

The four farms differ with regard to spatial extent and intensity of maize cultivation. Figure 3 shows the number of fields with maize cultivation from 2002 to 2007. The variation of the number of maize fields between the years is on farm A and D higher than on farm B and C. Figure 4 shows maize cultivation frequency on the same field. Between 2002 and 2007 most fields were cultivated only once with maize. Few fields were cultivated more often (2 to 4 years) with maize.

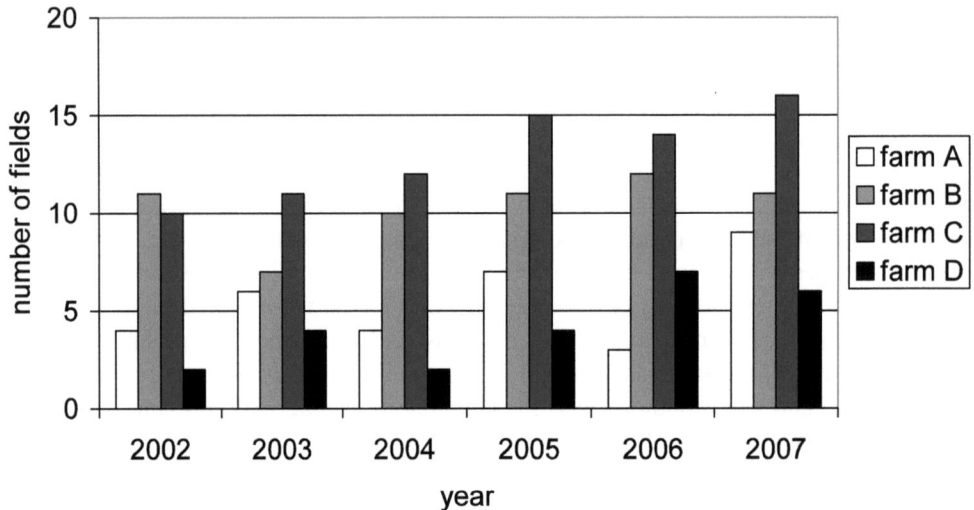

Fig. 3: Fields with maize cultivation in the four farms.

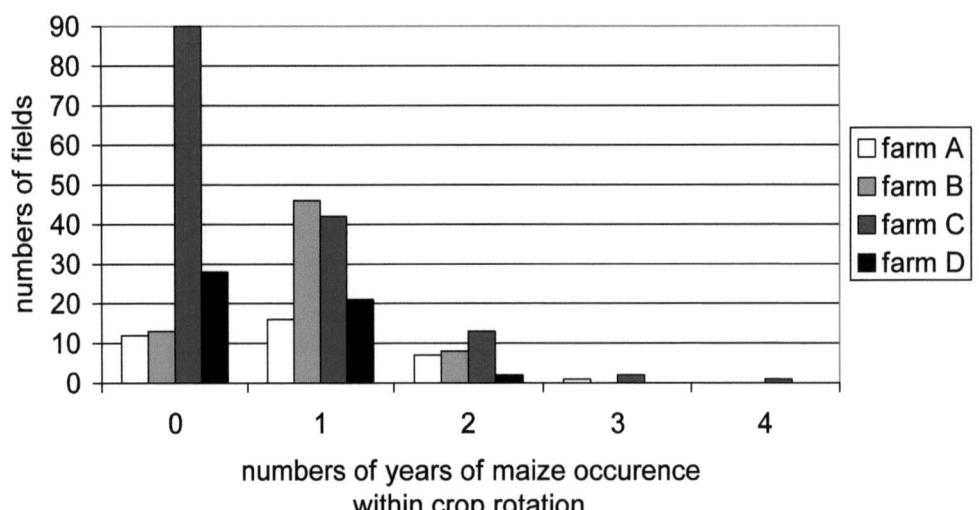

Fig. 4: Maize cultivation frequency on the same fields.

Tab. 1: Land use on the farm areas plus their neighbouring areas (buffers of 50 m / 200 m / 1000 m): percentage of the CLC 2000 classes (CLC 1: urban areas, CLC 2.1 and CLC 2.4: agricultural areas, CLC 2.2 and CLC 2.3: permanent crops, CLC 3: forest and seminatural areas, CLC 4: wetlands, CLC 5: water bodies).

Farm	urban areas	agricultural areas	permanent crops	forest and seminatural areas	wetlands	water bodies
---50 m----						
A	2	88	7	4	-	-
B	2	98	0.4	0.4	-	-
C	2	91	7	0.5	-	-
D	1	98	0.2	0.4	0.4	-
---200 m---						
A	3	78	8	11	-	-
B	5	92	1	2	-	0.02
C	3	89	7	1	-	-
D	2	95	0.4	1	1	-
---1000 m---						
A	2	59	6	30	0.01	3
B	6	87	2	5	-	0.4
C	4	86	7	2	-	1
D	3	94	1	1	1	0.001

Discussion and conclusions

The aim of this paper was to present our methodical approach to analyse the spatio-temporal patterns and dynamics of maize cultivation as a prerequisite to assess the potential exposure of Bt-maize on farm fields and surrounding biotopes. Based on the outcome of this analysis we intend to develop on-farm GMO monitoring schemes. The pre-processing and preliminary analysis of our data shows that the selected farms vary in their size, in extent and frequency of maize cultivation (Figures 1, 3 and 4) and in the types of surrounding biotopes (Table 1).

As a result of our analysis we expect to distinguish different categories and/or frequencies of exposition of fields and their surrounding biotopes to Bt-maize. We will possibly be able to indicate regular patterns as to how on-farm and surrounding GMO monitoring areas can be determined on the farm scale.

Our approach we consider as a case study within the study area of Märkisch Oderland and the four typical maize cultivating farms. Farm sizes, as well as extent and frequency of maize cultivation and types of surrounding biotopes vary throughout larger areas such as Brandenburg, Germany or Europe. Therefore the results of our analysis will not be directly transferable. However, we expect that once the data pre-processing procedures and the analysis of spatio-temporal patterns and dynamics of maize cultivation (Figure 2) are applied as presented above this methodical approach is transferable to larger areas.

GMO monitoring to be representative at larger scales requires including more spatial information such as GM crop cultivation patterns, ecoregion maps and existing environmental observation programmes (Graef et al. 2005). Developing on-farm GMO monitoring schemes should therefore be regarded as one important component to a representative larger scale monitoring approach.

References

Bossard M., Feranec J., Otahel J., Steenmans C. (2000) The revised and supplemented Corine land cover nomenclature. European Environment Agency, Copenhagen.

BVL (2010) Standortregister. http://apps2.bvl.bund.de/stareg_web/showflaechen.do.

European Community (2001) Directive/2001/18/EC of the European Parliament and of the Council. Officinal J EU Commun 2001/18/EC: 1–64.

Graef F., Züghart W., Hommel B., Heinrich U., Stachow U., Werner A. (2005) Methodological scheme for designing the monitoring of genetically modified crops at the regional scale. Environmental Monitoring and Assessment 111(1–3): 1–26.

LVLF Brandenburg (2005) Bericht des LVLF zur Überwachung des Anbaus von Bt-Mais MON810 im Jahr 2005. Potsdam, LVLF.

Meynen E., Schmidthüsen J., Gellert J., Neef E. Müller-Miny H., Schultze J. H. (Eds.) (1953–1962) Handbuch der naturräumlichen Gliederung Deutschlands. Bd. 1–9. Remagen, Bad Godesberg.

MLUV Brandenburg (2008) Gentechnik in Brandenburg. Bericht 2008. Potsdam, MLUV.

Statistische Ämter des Bundes und der Länder (2009) Statistik lokal. Daten für die Gemeinden, kreisfreien Städte und Kreise Deutschlands. Düsseldorf, IT.NRW.

Züghart W., Benzler A., Berhorn F., Sukopp U., Graef F. (2008) Determining indicators, methods and sites for monitoring potential adverse effects of genetically modified plants to the environment: The legal and conceptional framework for implementation. Euphytica 164(3): 845–852.

Chapter II

Dispersal and coexistence of GMO

Breckling, B. & Verhoeven, R. (2010) Implications of GM-Crop Cultivation at Large Spatial Scales. Theorie in der Ökologie 16. Frankfurt, Peter Lang.

Potential GM-maize cropping in Schleswig-Holstein I: Spatial heterogeneity of GM cultivation (Scenarios)

Christiane Eschenbach[a], Wilhelm Windhorst[a] & Andreas Rinker[b]
([a]Ecology Centre, University of Kiel, Germany; [b]Digsyland, Husby, Germany – ceschenbach@ecology.uni-kiel.de)

Introduction

On a regional scale, gene flow of maize (Zea mays, L.) cannot be precisely predicted for each individual field. However, large scale estimations on the efficiency of safety distances and the respective consequences on the landscape scale are urgently needed to identify potentials and expectable limitations caused by GM cultivation. Merging empirical small-scale gene flow data and a highly resolved spatial data base generated for Schleswig-Holstein (~ 15,000 km²) by connecting a GIS data base with statistical data on farm management and land use, it is possible to estimate gene flow on a regional scale assuming different scenarios of GM cropping intensities.

Methods and material

We present a data set for the federal state of Schleswig-Holstein (S-H, North Germany). In S-H (15,799 km², MLUR 2010, Figure 1a) 6,679 km² are cropped. The fields (n= 512,504) are typically small, as compared to other regions of Germany. Maize is a major crop in S-H, grown as silage maize. The fields are located mainly in the middle and eastern parts of S-H (Figure 1b). 2.9 % of the area of S-H are nature protection areas (Fauna-Flora-Habitat-Directive (FFH) and "Naturschutzgebiete" (NSG), Figure 1c). The number of farms amounts to 17,660 whereas 467 of them are managed organically (32,003 ha, 3.2 %) and preferentially located in the eastern parts with higher soil quality (Figure 1d). Up to now there were only two small sites with experimental releases or commercial cropping of GM-maize in S-H (2006–2008).

A spatial data base representing all farms was generated by connecting a GIS data base covering all 512,504 agricultural fields of S-H with statistical data on farm type, farm size and land use on the level of communities. Based on this information and information of satellite images it was possible to allocate maize cropping (24,710 fields, ~ 780 km² in 2000) and GM-maize cropping to specific fields and therewith to develop scenarios. The GIS application allowed the farm based grouping of GM-maize fields, taking farm based management decisions into account.

Scenario development

Scenario assumptions, as developed in the R&D project GeneRisk, were used for spatial assignment of GM crops and conventional crops. Isolation distances are considered in accordance with existing and discussed regulations according to the legal frame work: (1) The German Federal Act on Nature Conservation (BNatschG), and (2) Germany's Rules on Good Farming Practice in Producing Genetically Modified Crops (GenTPflEV): Farmers must leave 150 metres between fields of GM-maize and fields of conventional maize, and 300 metres between fields of GM-maize and fields of organic maize. The different scenarios combine different shares of GM-maize (10 %, 40 %, and 70 %) and different isolation distances to neighbouring conventional (150 m) and organic (300 m) non-GM-maize fields, and to nature conservation areas (1000 m). Additional scenarios refer to maize biomass production for bio fuel (plus 50 % maize).

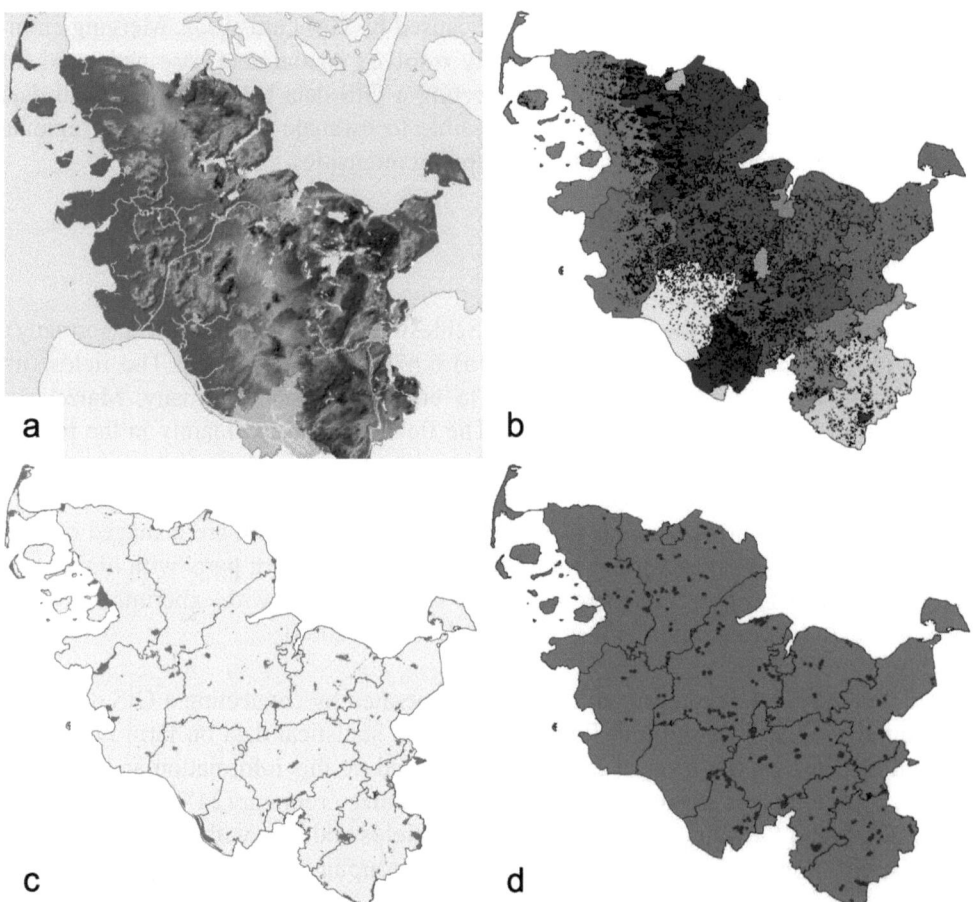

Fig. 1: The Situation in the Federal State of Schleswig-Holstein and its administrative districts (Landkreise): a) natural landscapes, b) maize cropping, c) nature protection, d) organic farming.

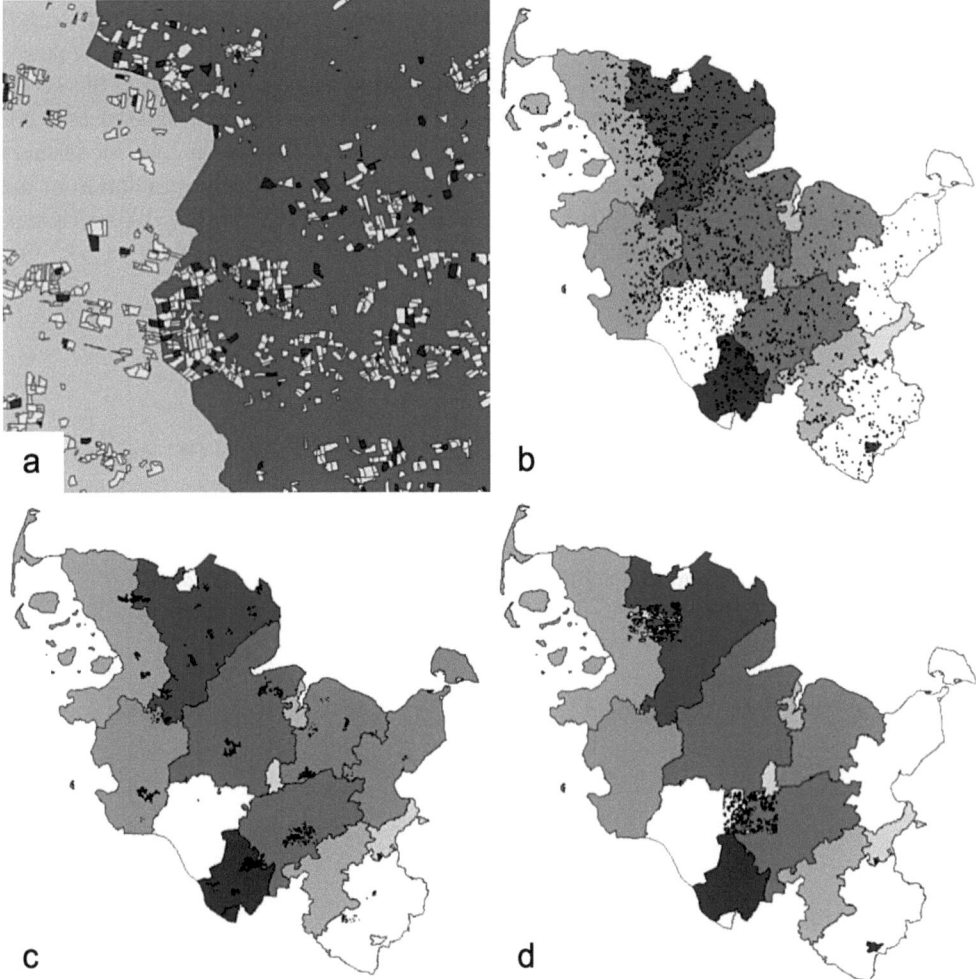

Fig. 2: Distribution of GM-maize fields for the 10 %-GM scenario in Schleswig-Holstein and its administrative districts (Landkreise): a) Detail: GM maize fields: dark polygons, other maize fields: light polygons, b) random without constraints, c) random with isolations distances to conventional non-GM-maize fields, to organic fields and to nature conservation areas, d) GM-maize fields clustered in two regions.

Results and discussion

In our study, these scenarios were applied to the conditions of S-H (Figure 2: 10 % GM-maize). The totally random distribution of GM-maize fields (Figure 2b) is unrealistic due to not randomly distributed constraints like organic farms. Spatial heterogeneity and sub-regional concentration of GM cultivation (Figure 2c) reflect existing and discussed constraints, and take management decisions of farmers into account. Considering the isolation distances according to GenTPflEV (150 and 300 metres) GM-maize cropping is limited to a share of 63 % in S-H. When calculations are carried out with isolation distances of 300 metres to neighbouring conventional non-GM-maize fields the limit is already reached by the 40 % scenario. Applying the rules of Good Farming Practice and assuming additionally buffer zones to nature conservation areas (1000 metres) the possible area available for GM-maize cropping is restricted to a share mimicked in the 10 %-GM maize scenario. Clustering the GM-maize fields (Figure 2d) is discussed as a co-existence measure to keep the adventitious presence of GM material low for farmers growing non-GM crops. The scenarios developed are the basis for the simulation of the spatial extent of gene flow using two different approaches, the MAMO (MAize MOdel, object oriented) simulation model and a GIS application (Eschenbach et al., this volume).

References

Eschenbach C., Breckling B., Rinker A. & Windhorst W. (2010) Potential GM-maize cropping in Schleswig-Holstein II: Model and GIS based approaches to estimate the gm-share in conventional maize yield. This volume.

GenTPflEV (2007): Verordnung über die gute fachliche Praxis bei der Erzeugung gentechnisch veränderter Pflanzen (Gentechnik-Pflanzenerzeugungsverordnung, 10.08.2007).

MLUR 2010: MLUR Ministerium für Landwirtschaft, Umwelt und ländliche Räume des Landes Schleswig-Holstein, www.schleswig-holstein.de.

Breckling, B. & Verhoeven, R. (2010) Implications of GM-Crop Cultivation at Large Spatial Scales.
Theorie in der Ökologie 16. Frankfurt, Peter Lang.

Potential GM-maize cropping in Schleswig-Holstein II: Model and GIS based approaches to estimate the GM-share in conventional maize yield

Christiane Eschenbach[a], Broder Breckling[b], Andreas Rinker[c] & Wilhelm Windhorst[a]
([a]Ecology Centre, University of Kiel; [b]University of Vechta; [c]Digsyland, Husby; Germany. – ceschenbach@ecology.uni-kiel.de)

Introduction

A large number of local field experiments is reported in the literature, in which gene flow of maize (*Zea mays*, L.) was measured. The results inform about single cases, however, it is not possible to draw direct conclusions for the regional scale without employing model calculations. For whole regions, gene flow cannot be precisely predicted for each individual field. Here, we present regional gene flow model calculations using a dispersal kernel in two different procedures. In addition to cross-pollination data, a spatial data base generated by connecting GIS data sets with statistical data on farm management and land use was required for the calculations.

Estimation of admixture in the harvest for different cultivation situations on the landscape or regional level can contribute to the analysis of co-existence implications. Co-existence refers to the ability of farmers to choose between conventional, organic or GM-based crop production systems (2003/556/EC). EU regulations have introduced a 0.9 % labelling threshold for the adventitious presence of GM material in non-GM products (2003/1829/EC). The ability to maintain different agricultural production systems ensures a high degree of consumer choice. Isolations distances are discussed and applied as measures to ensure co-existence (GenTPflEV 2007; Sanvido et al. 2007).

Methods

Our study is exemplified as a case study for the federal state of Schleswig-Holstein (S-H), located in the most northern part of Germany. Maize is a major crop in S-H, grown as silage maize. Up to now there have been only two small sites with experimental releases or commercial cropping of GM-maize in S-H (2006–2008). Therefore, in order to describe and quantify effects of potential GM maize cultivation we work with scenarios. Scenario assumptions were used for spatial assignment of GM and conventional crops (see Eschenbach et al., this volume, for further description). The scenarios illustrate different shares of GM maize. The GIS application allowed the farm based grouping of GMO fields. According to existing and discussed regulations the scenarios apply the isolation distances to neighbouring fields according to the rules of good agricultural practice in Germany (GM maize can be cultivated only if the next conventional maize field is more than 150 m away, and if the next organic maize field is no closer than 300 m). Spatial heterogeneity and sub-regional concentration of GM cultivation reflect these constraints. The scenarios developed are the basis for the simulation of the spatial extent of gene flow using two different approaches, the MaMo simulation model (Reuter et al 2008) and a GIS application.

As there are no wild relatives of maize in S-H and maize seeds or seedlings do not survive winter temperatures here, the major source of mixing is pollen flow. Pollen is produced in high abundance and maize is mainly wind-pollinated. The dispersal kernel is based on empirical data and was derived from a comprehensive literature study (Reuter et. al. 2008). The model MaMo uses a dispersal kernel in form of a table function as input. The model can be specified with a parameter for flowering synchronization. The duration of flowering period was considered to take 10 days for each field with standard deviation in starting the flowering of 1, 6, or 18 days chosen randomly for the different fields. The standard deviation of 1 day represents a scenario of full flowering synchrony in the region, while the 18 days standard deviation stretches the flowering phase beyond of what is usually expected.

With the GIS application the same dispersal function as in MAMO was used. A variation in time of flowering could not be considered. Crop cultivation geometry was considered in different ways in the two approaches, the MaMo model and the GIS application. MaMo uses a spatially explicit approach with a simplification of field geometries. Each field was represented by its centroid location and a weighting factor conforming to its size. Since this ignores situations with higher boundary/area relation due to complex field geometries, the approach tends to underestimate pollen transfer. The maximum cross-pollination distance between any two fields is set to 4500 m.

The GIS application as an alternative approach uses the real field geometry. A buffer zone of 500 m surrounds the GMO field in this case. Within this distance all maize fields are represented by a pattern of dots (representing squares of 10 x 10 m). For each dot the amount of pollen was calculated in dependency of the distance from edge of the GMO field(s) and the pollination percentage for the whole field was averaged.

Fig. 1: Results of a MaMo Scenario run (40 % GM-maize. Separation distances: GM to conventional maize 150 m, GM to organic maize 300 m).
Above: Scenario assignment of GM fields and conventional fields across Schleswig-Holstein. Light gray: conventional fields, black: GM fields.
Below: simulation result. The calculated cross-pollination is drawn as a bar upwards from each field centroid on a logarithmic scale (see scale in the legend right side for comparison). Conventional maize impurities occurring in GM fields are drawn light gray. GM impurities in conventional harvest are drawn black. It is apparent, that considerable impurities occur despite segregation distances.

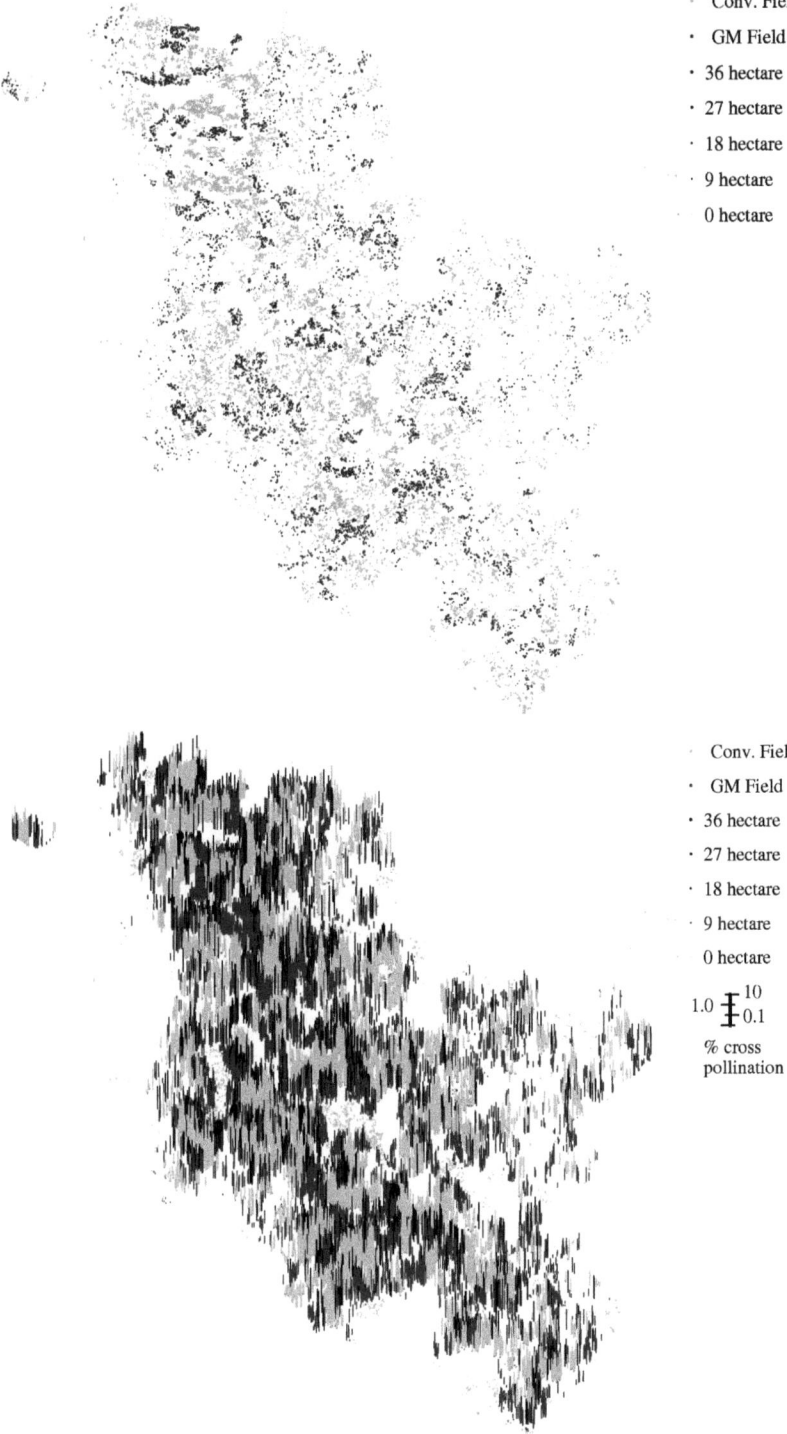

Results

Figure 1 shows exemplary results for a MaMo simulation run. As in the depicted crop assignment, the isolation distances in combination with the farm structure lead to a clustered assignment of GM cultivation.

The average % of GM impurities in conventional harvest was calculated according to the different scenarios (Tab. 1). Most interesting is the percentage of the conventional cultivation area for which GM impurities exceed the 0.9 % labelling threshold. For example, in case of the 10 % GM scenario the harvest of about 1.6 % of the conventional cultivation area is expected to exceed the threshold of 0.9 %. For the 40 % scenario the harvest of even 10 % of the non-GM maize area in Schleswig-Holstein might have to be labelled and sold as GM maize.

The results obtained by GIS application and the results of the MaMo-model show a considerable conformity. The values obtained by GIS application in all cases are in the range of the MaMo simulations exhibiting the different flowering synchronisations (cross-pollination is higher for a shorter flowering periods).

Tab. 1: Results of the scenario-calculations of 10 % and 40 % GM cultivation in Schleswig-Holstein. The GM field assignment considered the distance regulation of 150 m to conventional and 300 m to organic fields. The MaMo model was used with a maximum cross-pollination distance of 4500 m and different standard deviations in flowering synchrony (1 day, 6 days, 18 days), In MaMo, the crop geometry is simplified to the field centroid location and a weighting factor according to the field size.
The same scenarios were calculated with a GIS. This alternative approach took into account the field geometries on a 10 x 10 m grid base up to a maximum cross-pollination distance of 500 m around each single field.

	Scenario with 10 % GM cultivation				Scenario with 40 % GM cultivation			
	MaMo			GIS	MaMo			GIS
	1 d std. dev.	6 d std. dev.	18 d std dev.	–	1 d std. dev.	6 d std. dev.	18 d std dev.	–
% conventional cultivation area with GM impurities > 0.9 %	4.39	2.6	1.04	1.6	22.58	13.68	5.16	10.0
	Scenario with 10 % GM cultivation plus 50 % maize for bio-fuels				Scenario with 40 % GM cultivation plus 50 % maize for bio-fuels			
	MaMo			GIS	MaMo			GIS
	1 d std. dev.	6 d std. dev.	18 d std dev.	–	1 d std. dev.	6 d std. dev.	18 d std dev.	–
% conventional cultivation area with GM impurities > 0.9 %	6.71	3.89	1.47	1.8	24.51	15.46	6.2	16.9

Conclusions

Both methods, MaMo and GIS-application, appeared suitable to estimate geneflow at a regional scale and showed well comparable results. The cross-pollination results obtained with a GIS were always inside the range of variation obtained with the MaMo model that took flowering synchronisation effects into account. For all scenarios investigated it was demonstrated, that despite the application of separation distances a remarkable share of the conventional maize harvest in S-H would have to be labelled as GM maize because of impurities exceeding the labelling threshold of 0,9 %. It has to be noted, that the simulations do not take into account other important drivers or constraints like wind direction (e.g. Hoyle & Cresswell 2007). Also they did not account for any seed impurities or any other processes than cross-pollination, which may lead to additional increase of GM harvest in conventional maize cultivation.

References

2003/556/EC: Commission Recommendation of 23 July 2003 on guidelines for the development of national strategies and best practices to ensure the coexistence of genetically modified crops with conventional and organic farming.

2003/1829/EC: Regulation of the European Parliament and of the Council of 22 September 2003 on genetically modified food and feed (food and nutrition labelling).

GenTPflEV (2007) Verordnung über die gute fachliche Praxis bei der Erzeugung gentechnisch veränderter Pflanzen (Gentechnik-Pflanzenerzeugungsverordnung, 10.08.2007).

Hoyle M., Cresswell J.E. (2007) The effect of wind direction on cross-pollination in wind-pollinated GM crops. Ecological Applications 17(4): 1234–1243.

Sanvido O., Widmer F., Winzeler M., Streit B., Szerencsits E., Bigler F. (2007) Definition and feasibility of isolation distances for transgenic maize cultivation. Transgenic Research 17: 317–335.

Reuter H., Böckmann S., Breckling B. (2008) Analysing Cross-pollination studies in maize. In: Breckling B., Reuter H., Verhoeven R. (eds) Implications of GM-Crop Cultivation at Large Spatial Scales. Theorie in der Ökologie 14. Frankfurt, Peter Lang Verlag. 47–53.

Breckling, B. & Verhoeven, R. (2010) Implications of GM-Crop Cultivation at Large Spatial Scales.
Theorie in der Ökologie 16. Frankfurt, Peter Lang.

WebGIS as a tool for GMO monitoring support and for identification of potential coexistence problems due to GMO cultivation

Lukas Kleppin, Gunther Schmidt & Winfried Schröder
(Chair of Landscape Ecology, University of Vechta, Germany. –
lkleppin@iuw.uni-vechta.de)

Goal and background

The introduction of GMO (Genetically Modified Organisms) in agricultural ecosystems may cause adverse and irreversible impacts. Risk assessment on possible impacts mainly focuses on empirical studies with small spatial extent (e.g., small-scale field trials). Long-term monitoring of possible environmental negative effects, e.g., decrease of biodiversity, at landscape level is not established, yet, but indispensable (Henle et al. 2008; Schmeller & Henle 2008). Changes in agricultural practise have to be considered as well (Graef et al. 2007; Hails 2002). Insect-resistant Bt-maize is the only crop that was licensed for commercial cultivation in Germany (2005–2008). Thus, an assessment of interactions between genetically modified maize cultivation and conventional maize cultivation (coexistence) or nature conservation area is of special importance. Accordingly, the surveillance of GMO cultivation is an emerging field in environmental monitoring. In this context, WebGIS (Web-based Geoinformation Systems) may serve as an appropriate tool for planning, coexistence regulation and risk assessment support. The objective of this contribution is to exemplify how to support GMO monitoring in the federal state of Brandenburg by means of a WebGIS.

Material and methods

As being a convenient alternative to proprietary software, Open Source software was used for implementing the WebGIS on GMO cultivation. The WebGIS is built up by the UMN Mapserver in combination with the Apache HTTP-server (Hypertext Transfer Protocol) and the database management system PostgreSQL including the spatial extension PostGIS. Further, the WebGIS-Client Suite Mapbender by CCGIS provides the user interface.

Most of the monitoring and GMO related data for Brandenburg were provided by the federal environmental authorities. The compiled database contains data on nature conservation areas and information on the occurrence of the Corn Borer (*Ostrinia nubilalis*) from 2005 until 2007 as being the target organism for Bt-maize use. Additionally, the WebGIS provides information on locations and measured items of relevant monitoring networks in Brandenburg, e.g., on soil, ground and surface water as well as on biodiversity. These monitoring networks are relevant for GMO monitoring because the Bt toxin can be accumulated in soil and water and, thus, may harm non-target-organisms (NTOs) resulting in changes of biodiversity. Accordingly, it should be practical to enhance measurements by tests on Bt concentrations and/or by measuring related direct or indirect effects. Hence, the locations of established monitoring sites should be considered when designing and operating a GMO monitoring network. Moreover, the WebGIS application provides maps on land use patterns and on ecoregions as well as agricultural data on cultivation intensity of several crops at district level derived from agricultural surveys for the years 1999, 2003, and 2007.

The GMO WebGIS application

The structure of the WebGIS application on GMO, its design and basic GIS-tools are described by Kleppin et al. (2009). Innovations were introduced with respect to extended GIS-techniques and additional data which support planning and realisation of monitoring of GMO. In the following, the WebGIS application is described in detail to illustrate benefits of the WebGIS regarding monitoring and coexistence issues.

The GIS techniques described in the following enable detection of possible areas of conflict where GMO cultivation may be incompatible with defined protection goals for nature reserves. A buffer function was implemented which enables generating buffers around selected geo-objects like, for instance, Bt-maize fields. In Figure 1a, at first a certain Bt-maize field was selected and then a buffer zone of 2,000 m around this field was defined (irregularly circle around the light grey maize field in the centre). In the next step, the user extracts all conservation areas within the buffer zone by applying a clipping tool (dark grey bordered shapes). Additionally, each conservation area is linked to a query template to derive detailed information on the extracted area, i.e., a list of protected goods at the respective reserve (Figure 1b). The spatial analysis whether a Bt-maize field is located within or near a conservation area is important because in the protected area NTOs might be affected by toxins produced by Bt-maize crops or a change in biodiversity of the respective area might be induced. Hence, the analysis reveals whether and what kind of NTOs have to be considered for further investigations.

For GMO monitoring purposes, the WebGIS supports implementation and efficient operation by identifying locations of observation sites of related environmental monitoring networks. Figure 2 illustrates how to identify appropriate monitoring sites in the neighbourhood of a Bt-maize field by applying the GIS-tools 'buffer' and 'clip'. Moreover, information on measured parameters is available to assess whether the respective monitoring sites are eligible for GMO monitoring. Figure 2 shows a buffer zone with an extent of 1,000 m around the Bt maize field in its centre indicating relevant observation sites on soil and water (Figure 2a). Further, measured parameters observed at the monitoring sites can be requested by the user (Figure 2b). This example should have illustrated how WebGIS might support GMO monitoring before and during GMO cultivation.

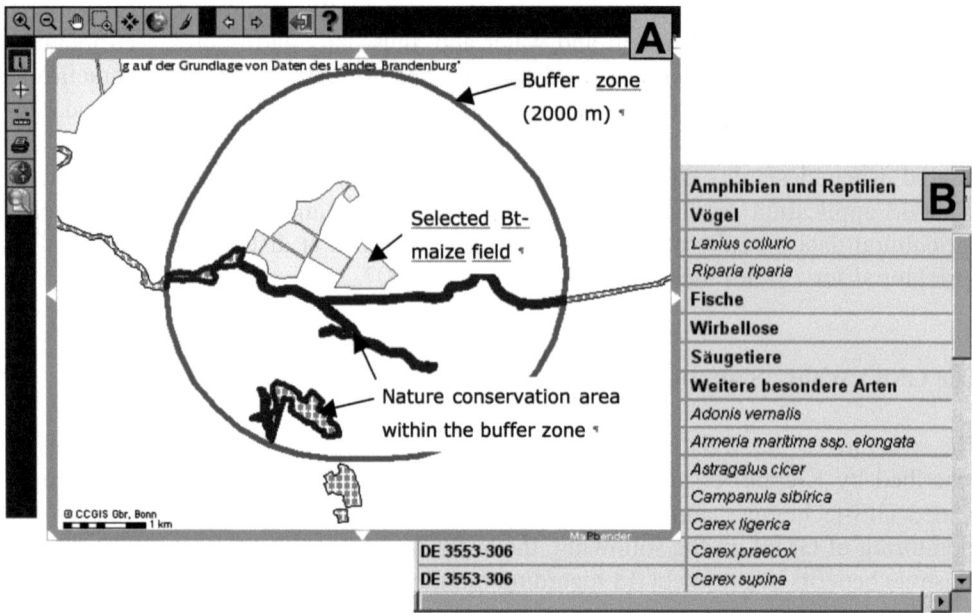

Fig. 1: GIS analysis: A) Bt-maize fields next to nature conservation area. B) List of protected species.

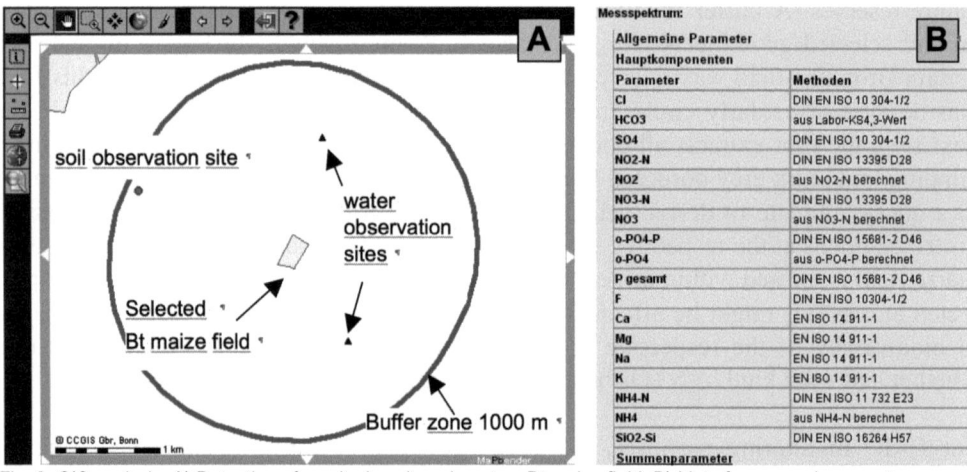

Fig. 2: GIS analysis: A) Detection of monitoring sites close to a Bt maize field. B) List of measured parameters.

Discussion

In Germany, only few web based GIS are available dealing with GMO. The "Risk Register Genetic Engineering Agriculture"[1] displays all Bt-maize fields in Brandenburg with the help of Google Maps. In comparison to the latter, the WebGIS GMO provides

---

1   http://www.risikoregister.de (Google Maps application).

GIS tools as well as additional environmental geodata and data on existing environmental monitoring networks. Compared with the German public GMO location register of the Federal Office of Consumer Protection and Food Safety (BVL)[2] the WebGIS GMO allows mapping of GMO sites and additional relevant geodata useful for supporting approval, monitoring, and risk assessment of GMO.

Outlook

At the moment, the GMO WebGIS application provides data important for GMO monitoring as well as tools for detecting regions where coexistence between GMO cultivation and conventional cultivation or nature potection might be difficult. Beyond this, the GMO WebGIS application will be complemented with additional data and web based functions. The WebGIS should also enable input, storage and analysis of the GMO monitoring data themselves. Following the WebGIS application 'MossMet' which was developed to support monitoring of 'Heavy Metals in Mosses' (Pesch et al. 2007), new extensions should be implemented to support GMO data surveys by the implementation of digital observation protocols. Furthermore, additional functions should be developed to allow a logical and spatial data queries and to export data and maps in established exchange formats. This should promote statistical evaluation of measurements and geodata. Finally, tools for upload of additional geodata by the local user should be made possible.

References

Graef F., Stachow U., Werner A., Schütte G. (2007) Agricultural practice changes with cultivating genetically modified herbicide-tolerant oilseed rape. Agricultural Systems 94(2), 111–118.
Hails R.S. (2002) Assessing the risks associated with new agricultural practices. Nature 418, 685–688.
Henle K., Alard D., Clitherow J., Cobb P., Firbank F, Kull T., McCracken D., Moritz R.F.A., Niemelä J., Rebane M., Wascher D., Watt A., Young J. (2008) Identifying and managing the conflicts between agriculture and biodiversity conservation in Europe—a review. Agric Ecosyst Environ 124, 60–71.
Kleppin L., Aden C., Schmidt G., Schröder W. (2009) WebGIS als Instrument für Planung und Monitoring des Anbaus von Bt-Mais. In: Strobl J., Blaschke T., Griesebner G. (Hrsg.): Angewandte Geoinformatik 2009. Wichmann, Heidelberg, 170–179.
Pesch R., Schmidt G., Schröder W., Aden C., Kleppin L., Holy M. (2007) Development, implementation and application of the WebGIS MossMet. In: Tochtermann K.; Scharl A. (Hrsg.) The Geospatial Web. How geo-browsers, social software and the Web 2.0 are shaping the network society. Springer London, 191–200.
Schmeller D.S., Henle K. (2008) Cultivation of genetically modified organisms: resource needs for monitoring adverse effects on biodiversity. Biodiversity and Conservation 17(14).

---

2   http://194.95.226.237/stareg_web/showflaechen.do (German GMO location register).

An African perspective of GM maize gene flow[1]

Chris Viljoen[a] & Lukeshni Chetty[b]

([a]GMO Testing Facility, Department of Haematology and Cell Biology, University of the Free State, Bloemfontein, South Africa; [b]South African National Biodiversity Institute, GMO Monitoring and Research, Applied Biodiversity Research, Pretoria, South Africa. – ViljoenCD @ufs.ac.za)

Abstract

South Africa is one of a few countries in Africa that has introduced genetically modified (GM) crops. South Africa has been growing 1st generation commercial GM crops since 1997 (Department of Agriculture 2005). In 2008, South Africa was ranked eighth in terms of global commercial GM production that included cotton, soybean, yellow and white maize (an important food staple) (James 2009). Since the introduction of GM crops, issues of coexistence and gene flow have not been considered a high priority by the agri-industry in South Africa. The reason for this is unknown but may stem from a lack of understanding the impacts thereof.

Gene flow from genetically modified crop to non-GM crop may have several consequences including: the development of resistance in insects for Bt crops; the contamination of landraces and loss of agro-biodiversity; loss of trade in processed and bulk grain commodities; the contamination of the food chain by experimental, industrial or pharmaceutical GM crops (Chilcutt & Tabashnik 2004; Reichman et al. 2006; Elbheri 2005.) Gene flow is a major contributor to commingling, and it occurs through pollen-mediated gene flow (PMGF) at farm level. Different out-crossing distances have been recorded for maize, using a variety of field trial designs under different environmental conditions, with the furthest distance being 650 m (Henry et al. 2003). However, these trials have usually been small plots and not on the scale of commercial farming. There is also no published data regarding the extent of PMGF for maize in South Africa, even after a decade of commercialization.

Thus aim of this study, conducted from 2005 to 2007, was to determine the extent of GM maize out-crossing under South African conditions in the context of commercial farming practice. Field trials were planted with a central plot of yellow GM maize (0.0576 Ha) surrounded by white non-GM maize (13.76 Ha), in two different geographic regions over two seasons with temporal and time isolation to surrounding commercial maize planting. Out-crossing from GM to non-GM maize was determined

---

[1] Extended abstract: A full paper is submitted to UWSF – Zeitschrift für Umweltchemie und Ökotoxikologie, Series: Implications of GMO-cultivation and monitoring. Springer-Verlag.

phenotypically, across 16 directional transects and confirmed genotypically using PCR. In this study, out-crossing occurred up to 300 m from the GM pollen source. The distance at which 0.1 % out-crossing occurred, ranged from 33 to 53 m, for 0.01 % out-crossing it was 113 m to 177 m, and for 0.001 % out-crossing it was 396 m to 596 m. Based on these data, an isolation distance of approximately 1295 m to 2009 m is required to contain out-crossing to a level of 0.0001 % given the parameters of the current study. However, these data also suggest that although the a level of 1.0 % out-crossing was achieved at approximately 20 to 25 m (with a GM pollen load from 0.06 Ha), the impact of pollen load from a typical Bt field in South Africa of 100 Ha will increase the required isolation distance considerably.

This study has shown that the current isolation distance used in seed production is not sufficient for preventing GM contamination, especially for industrial and pharmaceutical maize production. This study is a first for South Africa and it is hoped will contribute to the effective management of GM crops.

References

Chilcutt C.F., Tabashnik B.E. (2004) Contamination of refuges by Bacillus thuringiensis toxin genes from transgenic maize. Proceedings of the National Academy of Sciences: 101 (20): 7526–7529.

Department of Agriculture (2005a) Genetically Modified Organisms Act, 1997: Annual Report 2004/2005. www.nda.agric.za (accessed at 15 February 2007).

Elbehri A. (2005) Biopharming and the Food System: Examining the Potential Benefits and Risks. AgBioForum 8 (1): 18–25.

Henry C., Morgan D., Weekes R., Daniels R., Boffey C. (2003) Farm scale evaluations of GM crops: monitoring gene flow from GM crops to non-GM equivalent crops in the vicinity. Department for Environment Food and Rural Affairs, United Kingdom. 1–25.

James C. (2009) ISAAA Report on Global Status of Biotech/GM Crops. ISAAA Brief No.39. ISAAA: Ithaca, NY.

Reichman J.R., Watrud L.S., Lee E.H., Burdick C.A., Bollman M.A., Storm M.J., King G.A., Mallory-Smith C. (2006) Establishment of transgenic herbicide-resistant creeping bentgrass (Agrostis stolonifera L.) in nonagronomic habitats. Molecular Ecology 15: 4243–4255.

Breckling, B. & Verhoeven, R. (2010) Implications of GM-Crop Cultivation at Large Spatial Scales.
Theorie in der Ökologie 16. Frankfurt, Peter Lang.

Proposal for large-scale regional monitoring of genetically modified maize crops in small-scale agricultural systems in Africa[1]

Denis Worlanyo Aheto
(School of Biological Sciences, University of Cape Coast, Ghana. –
worlaheden@yahoo.com)

Purpose

This paper attempts to provide innovative guidance for future development of large-scale GMO monitoring schemes for small-scale agriculture in Africa. It is intended to contribute to studies focusing on biosafety analysis in Africa, which at present are widely lacking. The limitation of scientific data in this field is of grave concern owing to the uncertainties of the development of GM crops in the region. Presently, there are serious arguments that compel the need to monitor and thus to validate pre-market risk assessment measures and testing of transgenic crops. Low income African nations are particularly faced with a problematic issue since genetically modified varieties relevant on the world market have been developed, tested and notified for conditions that differ in climatic conditions, agricultural practice, biota and consumer preferences. Weak administrative competencies coupled with the different forms of seed procurement and use further compound the issue.

Using Ghana as an example, the paper will address biosafety issues by focusing on a specific sector of agricultural food production in urban areas of Africa. This type of low income investment contributes to subsistence as well as generates some small income for most inhabitants and migrant people. Though being a dynamic and rapidly expanding sector of present African agriculture, it is widely unrecognised by agro-ecological research. This sector of food production has so far not been addressed in biosafety research.

Due to the dynamic social context and the proximity to metropolitan areas, including the ports and harbours, it can be expected that access to incoming foreign technologies would make genetically modified crops readily available to farming systems in peri-urban areas and beyond, either directly in the form of commercial seeds or indirectly as re-sown fractions of imported food subsidies. The Cartagena Protocol on Biosafety (2000) confirms the sovereign rights of states to specify their own legal and administrative procedures on an informed scientific basis.

---

[1] Extended abstract: A full paper is submitted to UWSF – Zeitschrift für Umweltchemie und Ökotoxikologie, Series: Implications of GMO-cultivation and monitoring. Springer-Verlag.

In Africa, regulations as well as enforcement in this field are still widely lacking. While the African Union has started to operate an Africa-wide biosafety framework project, the majority of African states have only made the very first steps in assigning administrative competencies. The paper will give an overview of the feasibility of different methods covering a spectrum of spatial analysis, socioeconomics, and ecological modelling applications. It will depart from European biosafety research by focusing on relevant aspects of farming practices in Africa covering the following:
- Highlight aspects including differences in receiving environments;
- Provide information on the demographic, gender issues and livelihood conditions;
- Provide data on the agro-structure, seed procurement and technology access;
- Suggest protocols for improving the design of monitoring and regulatory schemes;
- Outline possible implications of cultivation of GM crops in Ghana and address relevance of data for other comparable countries.

Methods

Ground-based mapping of maize fields was done using a GPS receiver to determine the spatial configuration of the fields in 25 km^2 region and analyzed using ArcGIS. These protocols were originally developed by Gertrud Menzel in 2006. Cross-pollination estimates between GM and conventional fields were derived by applying an ecological simulation model according to Reuter et al. (2008a) and Reuter et al. (2008b). The model uses simplified geometric structures of fields and therefore allows to apply for larger regions. The model was developed within the Social Ecology Call of the Gene-Risk Project for European conditions (Gene Risk 2005–2010). Parametization of the model was done to accommodate information on the Ghanaian conditions such as crop field locations, variability in the sowing dates, the vegetative period, duration of pollination and maximum distance of dispersal. Isolation distances from field neighbours were done using a computer programme. A socioeconomic implication assessment of coexistence of GM and conventional cropping is also made through the administration of questionnaires among a wider segment of farmer population.

Analysis and conclusions

The data showed that the given cropping density provides optimal conditions for transgene spread, potentially limiting the possibility for coexistence between GM and non-GM fields. The data revealed on the average about 60 fields within a nearest distance of 100 meters, and cropping density of 56 fields per square kilometre. For a single GM field used as the most minimum scenario, an average GM crosspollination rate of 0.12 % of conventional fields would have to be expected. The data also allows to estimate the effects resulting from not only for single GM central fields but also to consider the effects of random processes, how the entry of GM seeds obtained from food and seed markets or eventually from previous harvests could influence cross-pollination in conventional fields to a large extent.

It is concluded that the size of a recipient field, its location and the distance to a GM field are important parameters to estimate the probability of transgene introgression. The data indicates that GM varieties if introduced could remain in cultivation even if their admission has expired or has been retracted. This would be undesirable and could cause long-term, undesirable stacked combination of transgenes which cannot be tested according to eventual combinatory effects. The data including the implied challenges coincides with what was reported in Brazil by Alves and Ogliari in 2008. The initial results confirm that the introduction of GM varieties could pose major risks to farmer livelihoods and the conservation of maize genetic resources with dire consequences for the African food export sector.

While these findings may be regarded as preliminary, a wider application of the field protocols and the model is proposed, especially for low income African nations or other comparable countries in the developing world to allow for fuller consideration of the implications of GM cultivation in small-scale agriculture.

References

Alves A.C.A.C., Ogliari J. (2008) Challenges for coexistence in small-scale farming: the case of maize in Brazil In: Breckling B., Reuter H. Verhoeven R. (eds.) Implications of GM Crop Cultivation at Large Spatial Scales. 134–139.

Cartagena Protocol on Biosafety (2000) (Under the Convention on Biological Diversity, CBD). http://www.biodiv.org/biosafety/protocol.shtml.

GeneRisk (2005–2010) Ecological and Economic Analyses about Co-Existence of Agriculture with and without Genetically Modified Plants:
http://www.sozial-oekologische-forschung.org/en/692.php.

Menzel G. (2006) Verbreitungsdynamik und Auskreuzungspotenzial von Brassica napus L. (Raps) im Großraum Bremen. Basiserhebung zum Monitoring von Umweltwirkungen transgener Kulturpflanzen. Dissertation University of Bremen.

Reuter H., Menzel G., Pehlke G., Breckling B. (2008a) Hazard Mitigation or Mitigation Hazard? Would genetically modified dwarfed oilseed rape (Brassica napus L.) increase feral survival? Environmental Science and Pollution Research 15 (7): 529-535.

Reuter H, Böckmann S, Breckling B. (2008b) Analysing cross-pollination studies in maize. In: Breckling B., Reuter H. Verhoeven R. (eds.) Implications of GM-Crop Cultivation at Large Spatial Scales. Theorie in der Ökologie 14. Frankfurt, Peter Lang, 47–53.

Influence of GM-crop cultivation on local apiculture and floral environment

Peter Wagner
(member of DIB – German beekeeping association. – wagner-pj@arcor.de)

The view of this article is not the communication of specific scientific findings but an issue of stakeholder involvement. GMO implications are largely discussed from an agricultural perspective, leaving out other stakeholders. As a local beekeeper I was invited to describe perspectives and preoccupations that many colleagues share concerning the effects which new developments in agriculture may have on apiculture.

Present situation

In Germany, there are about 90,000 beekeepers taking care of appr. one million bee colonies and, hence, contributing to the benefit of the local flora and biodiversity through pollination services. Most of them are hobby beekeepers maintaining in average 2.8 colonies per km² (DIB 2009).

The most obvious effect of beekeeping is the production of honey, but the most important effect is pollination. The economic value of pollination – which beekeepers and bees manage "by the way" and free of costs – is estimated to be about 10 to 15 times as high as the value of total honey crop. Thus, the honey-bee classifies as number three of all the productive livestock in Central Europe in economic importance (Tautz 2007). In global terms: In 2005, the worth of pollination equaled about 150 billion $, nearly 10 % of the value of total world food production (UFZ 2008).

Looking at the history of apiculture we can see different kinds of relationships: between honey-bees and flowering plants – nectar for pollination; between man and honey-bee – honey for shelter and maintenance; between beekeepers and farmers – rich harvest for both of them. Each of these relationships is characterised by a classical win-win-situation.

During the last decades beekeeping has been continuously declining due to adverse impacts like new diseases and parasites and the application of insecticides in agriculture. Monocultures are another reason. As a result, some regions (SZ 2010) are already free of honey-bees causing the according negative consequences for the fertility of flora. Some beekeepers have to feed their bees even in summertime – usually the main foraging period – to keep them from starving.

Some facts about honey-bees

By foraging in the surroundings of their hives, bee's honey reflects the environment like a fingerprint. Variety of flora and indirectly even the weather conditions of the last weeks can be derived as well as the range of pesticides or even nuclear fallout. Honey is still considered as relatively clean food and remedy but we cannot expect it to remain cleaner than the environment we offer to the bees.

The average foraging range of a bee-colony is about 3 km. This equals a covered area of nearly 30 km² per colony. In Germany, we have to expect 80 other colonies within the same area which means up to 4 million individuals in summer (more than 50,000 bees per colony, LdBK 1987). If GMO are grown in the same area it is obvious that sooner or later bees will cause hybridisation between GMO and non-target-plants as pollination is one of their main tasks. The introduction of buffer areas with a range of 150 or 300 m around GMO fields is just as futile as the instruction to cut off the tassels of the GMO.

No matter whether honey is looked upon as food of herbal or livestock origin, no matter whether or not tolerance levels are going to be defined by legislation, there is no doubt that smaller or higher amounts of GM-pollen will occur in honey that is collected within 3 km distance of a field where GM-plants are grown (LUA 2009).

Which are the consequences we will have to face?

For the farmers
Transfer of GM-pollen on the fields of traditionally working farmers may cause problems – for organic farming it will mean loss of the basis of their income since even traces of GMOs are forbidden in organic products by own obligation.

Often GMOs are developed to resist herbicide treatment or to produce insecticides in all parts of the plant. Through cross breeding, wild relatives may evolve expressing the same traits which makes weed control quite difficult. Plants being insect resistant during all of their lifetime may get an unwanted advantage compared to others due to the lack of natural enemies. They might tend to dominate the flora. – Is there an advantage for the farmers in the long term?

For the beekeepers
About 80 % of the customers refrain from buying / eating honey and other food containing parts of GMOs. In this context, it is irrelevant whether pollen grains are looked upon to be organisms of their own or not. Pure pollen as protein rich food will not be bought any more. To illustrate the implications: In Germany, honey containing traces of pollen of genetically modified maize MON 810 was classified not to be marketable. (VerwG 2007) (in Germany, MON 810 is only approved for feed – but not for human use. Honey is not allowed to have ingredients that are not suitable for human consumption). In this case, the respective beekeeper decided to depose the complete harvest as

toxic waste (taz 2008). To avoid such incidents in future beekeepers are now required to avoid the vicinity of GMO fields. But where to go, where will we find an area of 30 km² free of GM crops if GM agriculture expands? And secondly, in case of removing the bees the necessary and desired pollination will not take place. Where is the advantage for agriculture in the long term?

For the environment in general
The use of total herbicides and the cultivation of herbicide-resistant monocultures will definitely lead to a loss of biodiversity in the respective area. Existing plant communities will be destroyed. The honey of this "desert" (in bee's view) will become uniform and tasteless and will loose its flavour originating from the variety of different blossoms. The area will become less appealing for bees and beekeepers. Maybe they will withdraw or even quit beekeeping as contaminated honey cannot be sold anyway. The result will always be a decrease of pollination with negative consequences for the cultivation of fruits and crops in quantity and quality. Where is the advantage for agriculture in the long term?

Outlook

The legal implications of GM crop cultivation cannot be evaluated so far. Who will be held liable for cross-pollination? Who will compensate beekeepers and organic farmers for loss of income? Freedom of occupation is guaranteed by constitution – is this meant only for agro-chemical companies? Legal proceedings are inevitable but the basic issue cannot be solved. The beauty and floral variety of a whole continent (Europe) must not become a victim of short-term economic interests.

The position of the apiarists is well-defined and uncompromising (DIB 2010). Once set free, there is no way back for GMOs. The current win-win-situation between farmers and beekeepers will turn to its opposite. To avoid this, there is only one possible answer: Zero tolerance – coexistence is not practicable and therefore impossible.

References

DIB (2009) Bericht über die Tätigkeit des Deutschen Imkerbundes 2008/2009.
DIB(2010) Positionspapier des Deutschen Imkerbundes zur Agrogentechnik vom July 11, 2010.
LdBK (1987) Lexikon der Bienenkunde. München, Ehrenwirth Verlag.
LUA (2009) Landesumweltamt Brandenburg, GVO-Pollenmonitoring Ruhlsdorfer Bruch, Heft 110/2009.
SZ (2010) Manfred Hederer, Präsident des deutschen Berufs- und Erwerbsimkerbundes. Süddeutsche Zeitung, Issue of April 15, 2010.
Tautz, J. (2007) Phänomen Honigbiene. Heidelberg, Spektrum Akademischer Verlag.
taz (2008) die tageszeitung, Issue of October 2, 2008.
UFZ (2008) Helmholtz Zentrum für Umweltforschung, Leipzig, September 16, 2008.
VerwG (2007) Verwaltungsgericht Augsburg, May 4, 2007, Az.: Au 7 E 07.259.

Breckling, B. & Verhoeven, R. (2010) Implications of GM-Crop Cultivation at Large Spatial Scales. Theorie in der Ökologie 16. Frankfurt, Peter Lang.

Effect of two different gap crops on pollen-mediated gene flow in maize

Maren Langhof, Bernd Hommel, Alexandra Hüsken, Aldona Jarzmik, Joachim Schiemann, Peter Wehling & Gerhard Rühl
(Julius Kühn-Institut, Federal Research Centre for Cultivated Plants, Quedlinburg, Germany. – maren.langhof@jki.bund.de)

Introduction

One of the most adopted measures to assure coexistence of genetically modified (GM) maize and conventional or organic maize is an isolation distance between GM and non-GM fields. This space between fields may be planted with several crops and it is thought that crop types used as "gap crops" might differ in their effect on pollen-mediated gene flow (Devos et al. 2005; Langhof et al. 2008). In an earlier study we compared tall sunflower versus short clover-grass as gap crop; it was shown that pollen-mediated gene flow was not differently affected by these crop types (Langhof et al. 2008). In the current work the effect of barley stubble and clover-grass as gap crop on pollen-mediated gene flow in maize was investigated in order to prove if ascending air currents due to the heat up of the area grown with short and dry barley stubble enhance pollen-mediated gene flow above this crop surface in comparison to the area with short and wet clover-grass crop. Field experiments were conducted within the framework of the "Federal Research Programme on Coexistence" initiated by the German Federal Ministry of Food, Agriculture and Consumer Protection. In a three-year study (2007–2009) we tested these gap crops at two sites in northern Germany.

Materials and methods

Gene flow was investigated using a non-transgenic kernel-colour test system based on the dominant inheritance of the yellow-kernel colour over white kernel-colour. A field-in-field design was used, i.e. the pollen donor was surrounded by the pollen recipient. A 25 m wide strip of either clover-grass or barley was planted between donor and recipient maize (Figure 1). Gap crops were cut before maize flowering to allow the comparison of a short and dry (barley stubble) with a short and wet (clover-grass) crop. During the whole maize flowering period, meteorological data were recorded by on-site weather stations by the German National Meteorological Service. At each site flowering stages of both maize pollen donor and recipient were recorded at several sampling points within the donor and the recipient plots during the entire flowering period, approximately from mid-July to mid-August. Numbers of yellow kernels developing on white-kernel maize ears were counted, and the percentage of yellow kernels in relation to the mean total kernel number of a white-kernel maize ear was calculated.

Results and discussion

Results show that outcrossing was not influenced by the type of gap crop growing between donor and recipient maize plot. Overall, downwind outcrossing rates did not differ significantly between the short and wet clover-grass and the short and dry barley stubble gap crop (Table 1). Across all sites and years outcrossing was highest downwind from the donor, which was expected since maize is a wind pollinated crop. In the downwind direction levels of gene flow were highest in the first donor-facing row of the field edge (Table 1). Reduced pollen competition between incoming donor pollen and local recipient pollen as well as the short distance from the donor field are the main reasons for these "edge effects" (Langhof et al. 2008). Donor and recipient maize plants flowered synchronously at each site. During this three-year study wind speed and direction varied greatly between years and sites (data not shown), demonstrating the need for replications in field studies on pollen-mediated gene flow.

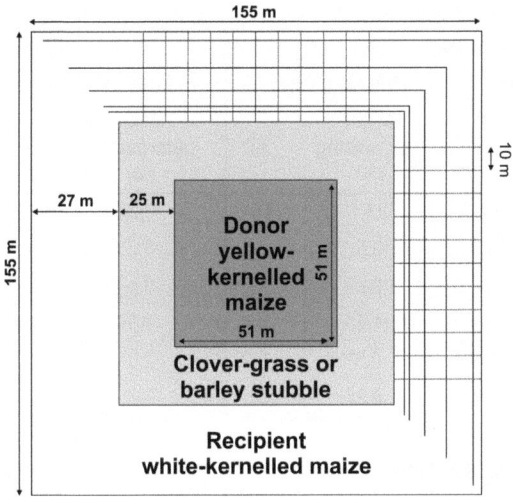

Fig. 1: Layout of the experimental field design established in 2007–2009 at Dahnsdorf and Braunschweig (2009 at Lucklum). Intercept points indicate maize ear sampling points at different locations within the white-kernelled recipient maize.

References

Devos Y., Reheul D., De Schrivjer A. (2005) The co-existence between transgenic and non-transgenic maize in the European Union: a focus on pollen flow and cross-fertilization. Environmental Biosafety Research 4: 71–87.

Langhof M., Hommel B., Hüsken A., Schiemann J., Wehling P., Wilhelm R., Rühl G. (2008) Coexistence in maize: Do non-maize buffer zones reduce gene flow between maize fields? Crop Science 48: 305–316.

Tab. 1: Downwind outcrossing rates (mean per row and standard deviation, SD) at different recipient maize field depths in dependence on barley stubble or clover-grass as gap crop at the sites Braunschweig/Lucklum and Dahnsdorf from 2007–2009. To define downwind areas in recipient fields the mean wind direction was set as the bisecting line of a 90° angle, with the angular point sitting in the centre of the pollen donor fields. All data points falling within the range of this sector were designated downwind and used for the statistical analysis. Within a row, asterisks indicate significant difference at each site (t-test, Bonferroni corrected p <0.008) between outcrossing rates in different field depths with barley stubble and clover-grass as gap crop.

Field depth (m)	Braunschweig 2007				Dahnsdorf 2007			
	Barley stubble		Clover-grass		Barley stubble		Clover-grass	
	Mean outcrossing (%)	SD	Mean outcrossing (%)	SD	Mean outcrossing (%)	SD	Mean outcrossing (%)	SD
0.0	19.2	7.6	19.2	8.3	16.9	11.6	8.5	5.5
1.5	7.6	2.8	7.5	2.5	7.3	3.8	5.6	2.6
3.8	3.7	1.6	4.4	2.4	3.9	2.7	2.0	1.4
6.0	1.1	0.3	1.7	1.2	2.1	1.0	1.3	0.7
11.2	1.5	0.9	1.6	0.9	1.0	0.9	0.6	0.4
23.3	1.0	0.6	1.3	0.8	1.1	0.6	1.2	0.9

Field depth (m)	Braunschweig 2008				Dahnsdorf 2008			
	Barley stubble		Clover-grass		Barley stubble		Clover-grass	
	Mean outcrossing (%)	SD	Mean outcrossing (%)	SD	Mean outcrossing (%)	SD	Mean outcrossing (%)	SD
0.0	26.5	12.5	22.4	16.6	10.2	7.6	11.5	6.6
1.5	4.3	1.8	6.5	3.3	8.5	4.4	10.0	6.5
3.8	1.5	1.4	2.8	1.2	3.2	2.4	5.9	5.2
6.0	1.1	0.9	1.4	0.8	1.6	1.4	2.4	1.9
11.2	0.5	0.3	0.9*	0.2	1.0	0.7	1.5	1.1
23.3	0.7	0.7	0.9	0.4	0.7	0.6	2.4*	1.6

Field depth (m)	Lucklum 2009				Dahnsdorf 2009			
	Barley stubble		Clover-grass		Barley stubble		Clover-grass	
	Mean outcrossing (%)	SD	Mean outcrossing (%)	SD	Mean outcrossing (%)	SD	Mean outcrossing (%)	SD
0.0	17.8	4.4	23.2	5.3	4.2	4.8	5.4	4.3
1.5	3.0	0.9	5.8*	2.2	1.3	1.9	1.2	1.9
3.8	1.3	0.6	2.0	0.6	0.3	0.2	0.3	0.3
6.0	1.0	0.4	1.8*	0.7	0.2	0.1	0.2	0.2
11.2	1.7	1.2	1.5	0.6	0.2	0.3	0.2	0.2
23.3	1.0	0.5	1.0	0.4	0.1	0.1	0.1	0.1

Chapter III

Management and control of GM-crops at large spatial scales

Comparative ecological effects of GM and other innovations in maritime arable-grass production systems

Geoffrey Squire, Mark Young & Alison Karley
(Scottish Crop Research Institute, Invergowrie, Dundee, UK. –
Geoff.Squire@scri.ac.uk)

Introduction

Agriculture in Europe needs to be multifunctional, to offer a balance among several outcomes. These outcomes, or functions, include food security, the aesthetic value of landscape, farm profit, rural living standards and choice by producer and consumer. None of the outcomes should hold more importance, however, than the continued ecological 'health' of the production ecosystem. This paper outlines an approach to ecological health that is being applied systematically to a typical arable-grass production system in a maritime region of north-west Europe. The basis of the approach is a set of *indicators* that are used to probe the biophysical properties of the system and to test and select the best methods of managing fields to achieve the desired outcomes. The approach can accommodate the assessment of innovations such as a new GM crop, but directs focus from the innovation itself to the needs of the system.

The indicators are measurable variables used to assess the state of the system. They are not constant over time, but can be imagined as varying within 'safe' limits if the system is to continue. (This concept may be more readily appreciated in terms of the ecological collapse that occurs when safe limits are repeatedly or continuously breached.) Innovations such as a new crop or field practice therefore need to be assessed according to whether they will maintain the indicators within safe limits or move them back towards safe limits if they have already been exceeded.

Indicators of current and target states

The approach was developed in a major research programme on the arable-grass systems of eastern Scotland. It aimed to define the salient indicators, measure their present value and assess their safe limits. The information on indicators was drawn from many sources: long term statistics and databases, field experiments, observation networks, process-based knowledge from controlled environments and models of the main flows and pathways of energy and matter. These sources were complemented by extensive, new field studies to gain baseline data over ranges of geography and farming intensity.

The main categories of indicator so far defined are a) inputs and operations on the field, b) flows and stores of energy, carbon (C) and water, c) plant production and offtake, d) cycles of nitrogen (N), phosphorus (P) and other nutrients e) organisms including plants, microbes, invertebrates, f) soil biophysical properties, and h) system-scale properties (e.g., solar-fossil ratios, emergy). Examples of indicators in the category – *energy, carbon, water* – are total dry matter production (t ha^{-1}), yield and other offtake (t ha^{-1}), calorific content of offtake (J g^{-1}), C/N ratios of different compartments and seasonal solar energy capture (GJ m^{-2}). The aim is to develop a basic set of field-tested indicators that can be used in all assessments. Examples of progress to date are given by Hawes et al. (2010) and Hillier et al. (2009). The study is confirming that the main factors influencing the internal health of the field are nitrogen and phosphorus fertiliser, tillage and offtake (as they affect soil properties) and the weed-based arable food web.

Assessment and comparison of impacts

While the safe limits for the main indicators are presently being defined, the direction in which indicators need to move is apparent in many cases. Moreover, changes in land management that have recently occurred, such as that from spring to winter cropping over a period of 10 to 20 years after the 1980s, have become important reference points. The change to winter cropping had large or moderate effects in all categories of indicator. For example, it altered the timing of agronomic operations, increased agronomic inputs, and caused greater carbon capture, more offtake of carbon and plant nutrients and major shifts in species and function in the food web. The several-fold changes in these indicators caused by winter cropping can be considered as 'large' in this system.

In comparison, an example of a moderate or small change is the replacement of conventional, chemical weed management with that associated with GM herbicide tolerant (GMHT) crops of oilseed rape. In extensive field testing in the UK (see references in Hawes et al. 2009), GMHT cropping reduced some beneficial plant and invertebrate indicator-groups, but also may have restricted choice (by producers and consumers) because of the potential mixing of GM and non-GM harvests. GMHT cropping appears to have little effect on total primary production and on cycles of energy and matter in these production systems.

Innovations that have not been field-tested are still examined by reference to general knowledge on the resilience of indicators, data on a near comparator and simulations of stock-and-flow models for energy and matter. For example, if all potential improvements in nitrogen uptake and efficiency were realised in advanced barley varieties, they would affect nitrogen, carbon, energy and food webs but less strongly than they were affected by the change to winter cropping. The introduction of a starch-modified GM potato, for example, is predicted to have minor effects on one or other of the crop quality indicators, but no effect on other indicator groups compared to current non-GM potato.

Concluding remarks

The approach outlined here sets the producing ecosystem first, defines the safe or sustainable limits in which the main biophysical indicators may operate, and designs crops and practices that would satisfy functions such as choice, food security and profit, while allowing the indicators to remain within or at least move towards safe limits. If the approach is to be applied successfully, the main challenges for management are to use and lose less nitrogen and phosphorus and to reverse declines in soil functioning and food webs. Existing examples from changes occurring over the past 30 years provide reference to what may be classed as small, moderate and in some cases large impacts of innovation on the indicators. GM crops that provide resistance to pathogens or added value through quality traits, yet barely alter cycles of energy and nutrients, are likely to have no more than small (and then probably positive) impacts on the main ecological indicators. Questions on future GM crops may need detailed scrutiny and evaluation, but are likely to be overshadowed by the main challenges facing sustained production over the coming decades.

References

Hawes C., Squire G.R., Hallett P.D., Watson C., Young, M. (2010) Arable plant communities as indicators of farming practice. Agriculture, Ecosystems and Environment. In press (doi: 10.1016/j.agee.2010.03.010).

Hillier J., Hawes C., Squire G.R., Hilton A., Wale S., Smith P. (2009) The carbon footprints of food crop production. International Journal of Agricultural Sustainability 7: 107–118.

Breckling, B. & Verhoeven, R. (2010) Implications of GM-Crop Cultivation at Large Spatial Scales.
Theorie in der Ökologie 16. Frankfurt, Peter Lang.

Demography of feral oilseed rape over 11 years in an agricultural region[1]

Gillian Banks, Mark W. Young & Geoffrey R. Squire
(Scottish Crop Research Institute, Invergowrie, Dundee DD2 5DA UK. –
Gillian.Banks@scri.ac.uk)

Introduction

While the crop rapeseed has been grown for centuries, its current widespread occurrence as a feral plant in rural and urban areas in Europe follows a marked increase in its cropped area in the 1970s and 1980s. The ferals can exchange genes with crops and may have ecological effects, for example by interacting with other ruderal cruciferous plants and the animals of the arable food web. Ferals have received most attention as potential contributors of impurity in crops. If GM oilseed rape were grown commercially, some of its spilled seed would enter feral populations, which may later transmit pollen and seed to non-GM fields. To provide evidence on the persistence and spread of ferals, a demographic study area of around 500 km^{-2} was established in Tayside, UK in 1993. This paper considers the size of the feral population over an 11-year period during which the area of land sown with oilseed rape declined.

Methods

The method of conducting the 'road and foot' survey in the 500 km^2 Tayside Study Area has been described by Charters, Robinson & Squire (1999). Individual groups of flowering crucifers on waysides, field margins and waste ground were inspected and their species composition determined. Feral oilseed rape, which comprised most of the plants in these groups, was counted at flowering, and generally also at seeding. Their position was noted on a geographic information system (GIS). The locations of all flowering fields of oilseed rape were noted and also transferred to the GIS. The whole survey was carried out twice – during the periods of flowering of winter oilseed rape (typically May) and spring oilseed rape (June/July). The total flowering population of feral oilseed rape in any year was determined as the sum of the individuals in the groups. The total number of flowering crop plants was estimated from the area of fields sown with the crop and the typical stand density which for the purpose of the analysis here is taken to be 50 m^{-2}. Flower and seed numbers on the crop were taken from the standard data used as input to the model by Begg et al. (2006). The original survey was conducted over four consecutive years 1993–1996. In subsequent years, surveys of part

---

[1] Extended abstract: A full paper is in preparation.

of the area were done in order to check for the presence of ferals. Then in 2004, a full survey was repeated. Consistency of observation method and personnel was maintained between the first four years and 2004.

Results

The crops that could give rise to feral oilseed rape in this region are winter (autumn sown) oilseed rape, spring oilseed rape (both *Brassica napus* L.), forage brassicas that only flower if not harvested or used to feed stock, and some oilseed varieties of the turnip, *B. rapa* L. The total area of these crops reached a maximum in the early 1990s. Thereafter, the areas of spring oilseed rape, forage brassica and turnip oilseed declined, almost to nothing by 2004, while the area of winter oilseed rape declined to around 70 % of the value in 1993. As a fraction of the land area under survey, the crops declined from 5.8 % to 3.3 %. In contrast, the total number of feral 'populations' (discrete sites at which ferals were flowering) and the total number of flowering individuals increased over the period. The early flowering ferals (synchronous with winter crops) increased from a mean of 47 populations in the first two years to 306 in the final year. The late flowering (synchronous with spring crops) increased from 20 in the first year to 55 in the final. Visual inspection revealed that they increased generally throughout the area, not just in a few localities. To estimate the statistical significance of the change, a 2 x 2 km grid was placed over the area, the number of ferals populations counted in each and the means tested for difference. The rise each year from 1994 onwards was significant for winter ferals ($F = 0.001$), while the smaller change in spring ferals was significant between 1993 and the final two years but not the intervening years ($F = 0.026$).

The net increase in feral populations resulted from a combination of persistence at sites and spread to new sites. Gradually, more of the area was colonised by ferals as assessed by a decrease in the mean size of the non-overlapping 'polygon' that could be placed around each feral population. However, the mean field-to-feral distance increased, especially for spring ferals, where the mode increased from around 1000 m in 2003 to 4000 m in 2004. Some of the spring ferals in 2004 were more than 10 km from the nearest field.

Despite the increases in feral population number, the total contribution of ferals to the flowering oilseed rape in the region was very small throughout the period. The area of the crop in the middle of the period was 2210 ha. At the nominal stand density, the number of crop plants would be 1.8×10^9. The number of flowering ferals in that year was just less than 5000 individuals, so ferals comprised 0.00028 % of the total flowering oilseed rape.

Discussion

During a decade when the areas of potential source crops had declined, the ferals increased. The contribution of spilled seed from other sources, for example transport of harvested rapeseed through the region, cannot be ruled out, but there were no changes to the locations of the major processing or storage operations. Moreover, any through-transport of rapeseed is most unlikely to deposit seed on many of the minor roads in the region, and it was on these minor roads and waysides where much of the increase occurred.

Notably, the late ferals had become separated from spring crop varieties. The genetic basis of these late flowering ferals had not been confirmed by molecular markers, but the germination in the year of flowering and absence of a need for vernalisation points strongly to a spring genotype. These isolated ferals may be evolving independently of the introduction of new seed.

The contribution of ferals to impurity in crops, and hence to GM coexistence, appears to be negligible at current population densities. If all the ferals in this region were harvested along with the crops, (which is an impossibility), they would make no noticeable difference to the total of impurity caused by cross pollination between crops or by volunteer seed. The feral oilseed rape in Tayside nevertheless presents an interesting case that demonstrates the long term dynamics of crop-feral demography.

References

Begg G.S., Hockaday S., Mcnicol J.W., Askew M., Squire G.R. (2006) Modelling the persistence of volunteer oilseed rape (*Brassica napus*). Ecological Modelelling 198: 195–207.

Charters Y.M., Robertson A., Squire G.R. (1999) Investigation of feral oilseed rape populations: genetically modified organisms research report (No. 12). Department of the Environment, Transport and the Regions.
http://www.defra.gov.uk/environment/gm/research/reports.htm.

The Triffid case: A short résumé on the re-discovery of a de-registered GMO

Gunther Schmidt & Broder Breckling
(Chair of Landscape Ecology, University of Vechta, Germany –
gschmidt@iuw.uni-vechta.de)

Introduction

Biosafety testing was established to assure that only GMOs are present in the environment that appear manageable and remain under administrative control. Now, for the first time it became apparent that a GMO has escaped administrative oversight for many years. We document the background and development of this case to emphasise that, in particular, the availability of routine detection for any GMO (perhaps also experimental release) is important. Otherwise, if undesirable developments occur, there may be long delays before possible contamination is detected. The described case points to an existing deficit that requires improvement, in particular, with regard to boundary crossing transfer of GMO or GMO products.

Background

In the late 1980's a GM flax (linseed) variety (*Linum usitatissimum*) known as FP967 and later named "Triffid" was developed at the Crop Development Centre in Saskatoon, Saskatchewan (Canada) by Alan McHughen using public funds [1]. The name was chosen by the developer synonymous to a fiction story about an oil-producing but dangerous carnivorous and invasive plant. Triffid flax was transformed with genes conferring resistance to sulfonylurea herbicides triasulfuron and metsulfuron-methyl, and resistances to different antibiotics as selectable markers.

Triffid flax had been admitted in Canada and in the US only for a very short time in the late 1990's. It was not well accepted by the growers. They feared harvest purity problems because the crop was not admitted in Europe as the main export market (70 %) [2]. Hence, Triffid was de-registered in 2001 and it was believed that all known stocks had been identified and destroyed [3]. Years later, in 2009, traces of Triffid were discovered in Canadian flax imported to Germany and subsequently in a larger number of different countries in Europe (Figure 1) and world-wide (e.g. Japan) [4], [5]. First traces were found in samples taken for food inspection in the federal state of Baden-Württemberg in Southwest Germany. Triffid was discovered in 16 of 41 conventional linseed samples (39 %), with a proportion of 0.05 to 0.1 % [6].

Consequences

The GM flax variety FP967 (CDC Triffid) is not authorised for food or feed use in the EU, meaning that any food product or flax / linseed derivative analysed to be positive for FP967 is not marketable in the EU. Zero tolerance applies, i.e. there are no exemptions even for minor traces of not approved GM products [6]. The impurities caused a financial loss of several million dollars for Canadian exporters. Within a few days, cash bids for flax in Manitoba dropped by 32 % when the contamination became public. Most of Canada's 2009 harvest remained in storage [7]. Additional costs emerged through the development of a specific test [8].

How could Triffid become so wide dispersed and did not vanish after almost ten years of de-registration? It may be possible that demonstration samples sent by Mc Hughen to farmers may have been sown. The Canadian Grain Commission (CGC) examines details. It has now been determined that two varieties of flax are contaminated with Triffid at the breeder seed level. In March 2010, the Flax Council of Canada is reporting that it has found extremely low levels of Triffid in more breeder seed samples. Triffid has now been discovered in the flax varieties CDC Bethune, CDC Sorrel and CDC Sanctuary [9]. As a result, there is now a shift in the issue of producers using farm-saved seed. The Flax Council says farm-saved seed will be allowed, but under rigorous sampling and testing procedures, a complex and costly task (ca. $ 105 per test) but assessed to be mandatory for the survival of the industry [10]. It appears that, if enough samples and sub-samples were tested and if the tests are sensitive enough, Triffid contamination is found to be widespread. In many of these tests, Triffid is only present at one or two seeds per million, but it isn't zero as required for GMO without admission. Meanwhile, the CGC is undertaking a geographic study of existing flax stocks held in commercial positions throughout Canada. The intention is to determine the nature and location of the GMO material [11].

Conclusions

It will be very difficult if not impossible to completely eliminate Triffid from the seed supply. For that reason, liability is going to be an issue for anyone selling flax seed. It has to be expected flax seed buyers will have to subscribe a liability waiver to prevent seed producer from possible claims of compensation.

Biosafety considerations require that for any GMO that was considered for commercialisation tests are routinely available. For Triffid, such a test had to be developed after the contamination was detected using unspecific, generic methods. This is certainly not acceptable with regard to present standards and requires administrative improvement. The case also draws attention to the fact that for any other de-registered GMO long-term availability of detection methods is a pre-requisite for a biosafety survey. These efforts increase with the number of released GMO.

Alan McHughen changed from Saskatchewan to the University of California, Riverside [12]. In 2000, he became founding president of the International Society for Biosafety Research and currently serves as its treasurer [13].

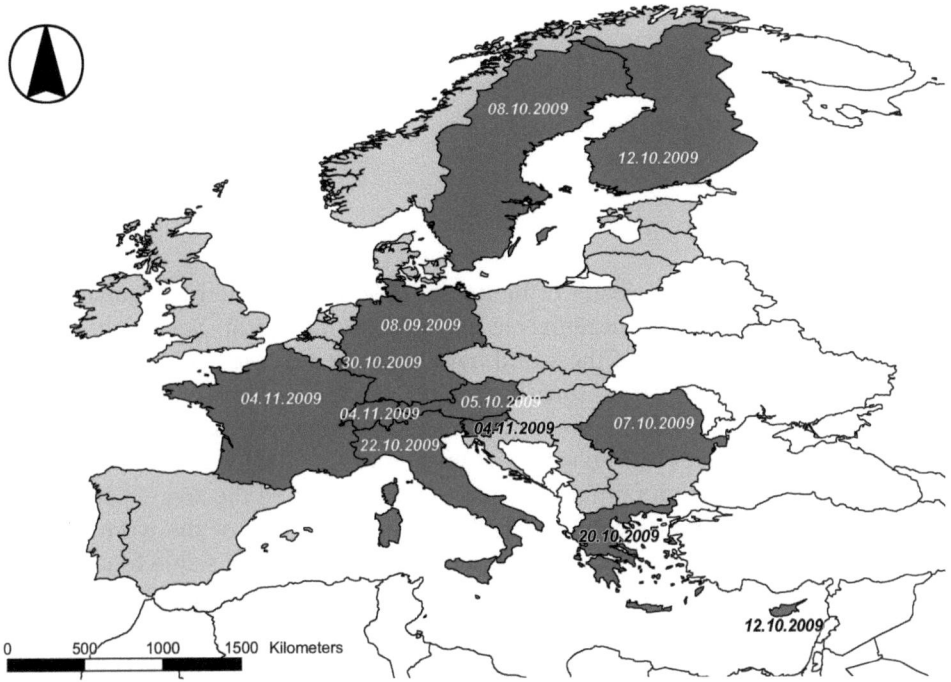

Fig. 1: Detection of Triffid flax in Europe; dates indicate first day of detection (dark grey = detected by laboratory tests, light grey = further distribution within Europe through commercial exchange of goods) (source:[4]).

Links
[1] http://www.spinprofiles.org/index.php/Alan_McHughen.
[2] http://www.cbc.ca/canada/manitoba/story/2010/01/20/mb-flax-triffid-manitoba.html.
[3] http://www.gmwatch.org/latest-listing/1-news-items/11557-flax-contamination-soars-to-28-countries.
[4] http://www.gmcontaminationregister.org/index.php?content=nw_detail1.
[5] http://ec.europa.eu/food/food/rapidalert/rasff_portal_database_en.htm (query tool).
[6] http://www.ua-bw.de/pub/beitrag.asp?subid=3&Thema_ID=17&ID=1206&Pdf=No.
[7] http://www.greenpeace.org/raw/content/international/press/reports/ge-contamination-devastates-ca.pdf.
[8] http://www.genetic-id.com/Pages_Link/Test-for-GM-Flax-FP967.aspx.
[9] http://www.prairiefarmandranch.com/cms/grain/triffid-contamination-continues.
[10] http://www.flaxcouncil.ca/files/web/MESSAGE%20TO%20PRODUCERSNov%2011,%202009.pdf.
[11] http://www.flaxcouncil.ca/files/web/GMO%20Flax%20Update%206%20October%202009.pdf.
[12] http://www.facultydirectory.ucr.edu/cgi-bin/pub/public_individual.pl?faculty=1912.
[13] http://extern.genius.de/web_access/intern/drupal/?q=node/100.

Monitoring maize diversity in Mexico for decision making[1]

Francisca Acevedo Gasman
(Comisión Nacional para el Conocimiento y Uso de la Biodiversidad (CONABIO), Tlalpan, Mexico. – facevedo@conabio.gob.mx)

Abstract

Mexico counts with a great wealth of biological diversity in its territory, and has been classified as one of the few megadiverse countries in the world. Several of the most economically relevant crops of the world have originated in Mexico, including beans, vanilla, cocoa, tomatoes, cotton, squashes, chiles and maize (see Biodiversidad Mexicana (2010) for further information). Although high tech extensive agriculture does exist in the northern states of Mexico, most maize farmers plant maize landraces which are cultivated depending on rainfall, and represent 86 % of all the area where maize is cultivated in Mexico (Bellon et al. 2009). While in some parts of the world, maize is used as food for feed mainly, in others it is still considered to be a staple food, probably the most important food crop in the world (Mercer & Wainwright 2008). Mexicans consume on average 336 grams of maize daily (last FAOSTAT report, 2005) translated into around 600 culinary dishes (Bourges 2002). Mexico as a country has also a very strong science force which has among other techniques opted for modern biotechnology developments whilst the big agricultural farmers that use extensive type farming are who are at its demand.

Mexico has a Biosafety Law in place since 2005, it regulates activities with GMOs and includes special attention to the concepts of "centers of origin and centers of genetic diversity" aiming towards the protection both of the species and where they are encountered. Among others, CONABIO must provide information to two competent authorities (the Ministries of Agriculture and Environment). These must determine and officially publish areas center of origin and where the known now a day genetic diversity of each species for which Mexico is center of origin and center of genetic diversity rests in order for these areas to be excluded from GM activities.

In the year 2006 CONABIO gave the relevant maize information it counted with including an analysis to these Ministries to determine these areas (CONABIO 2006). As a result, the two Ministries together with CIBIOGEM (an interministerial commission dedicated to biosafety issues) gave CONABIO between 2006 and 2007 1.5 million USD to promote the necessary studies to count mainly with recent information on genetic

---

[1] Extended abstract: A full paper is submitted to UWSF – Zeitschrift für Umweltchemie und Ökotoxikologie, Series: Implications of GMO-cultivation and monitoring. Springer-Verlag.

diversity of maize to be able to, with solid rock ground elements, determine the areas the law asks for. On centers of origin of maize, an analytical review of the published data has already been accomplished by a multidisciplinary group through this project initiative (Kato et al. 2009). CONABIO as of January 2010 has together with a group of institutions, almost completed the project solicited, and already counts with 21.290 out of the 23.167 registries compromised for maize landraces (91 %), 610 out of 453 registries for teocintles (130 %) and 537 out of 813 (66 %) for Tripsacum, the last two being wild relatives of maize and for which teocintle is thought to be the closest progenitor of maize.

Needs to fulfil the Biosafety Law requirements for decision making have triggered generation of new information in Mexico related to genetic diversity in maize. Results show mainly that diversity (maize landraces and wild relatives) is richer and wider spread than originally accounted for with the former data CONABIO counted with in its databases in 2006 for Mexico.

How will this information be taken into account by the two competent Ministries is yet to be seen, although much to do is foreseen if Mexico decides to go on with GM maize activities considering that the areas cover, from what we now know, most of the Mexican territory.

References

Bellon M.R., Barrientos-Priego A.F., Colunga-GarcíaMarín P., Perales H., Reyes Agüero J.A., Rosales Serna R., Zizumbo-Villareal D. (2009) Diversidad y conservación de recursos geneticos en plantas cultivadas, en Capital Natural de México, vol. II: Estado de la conservación y tendencias de cambio. CONABIO, México, pp. 355–382.
Biodiversidad Mexicana (2010) http://www.biodiversidad. gob.mx/genes/otrosCentros.html.
Bourges H. (2002) In: Alarcon-Segovia D., Bourges H. (eds) La Alimentación de los Mexicanos. El Colegio Nacional, México DF: 97–134.
FAO statistics, maize consumption in Mexico. http://faostat.fao.org/faostat (visisted January 2010).
CONABIO (2006) http://www.biodiversidad.gob.mx/genes/pdf/Doc_CdeOCdeDG.pdf.
Kato T.A. , Mapes C., Mera L.M., Serratos J.A., Bye R.A. (2009) Origen y diversificación del maíz: una revisión analítica. Universidad Autónoma de México, Comisión Nacional para el Conocimiento y Uso de la Biodiversidad. México, D.F. 116 pp. Download from: http://www.biodiversidad.gob.mx/genes/origenDiv.html.
Mercer K.L., Wainwright J.D. (2008) Gene flow from transgenic maize to landraces in Mexico: An analysis. Agriculture Ecosystems and Environment 123: 109–115.

Breckling, B. & Verhoeven, R. (2010) Implications of GM-Crop Cultivation at Large Spatial Scales.
Theorie in der Ökologie 16. Frankfurt, Peter Lang.

Study of maize fields and their surroundings in European regions regarding the suitability for coexistence of different maize cultivars

Sabine Prescher, Joachim Schiemann & Alexandra Hüsken
(Julius Kühn-Institute (JKI), Institute for Biosafety of Genetically Modified Plants, Quedlinburg, Germany. – sabine.prescher@jki.bund.de)

Abstract

Fields in four European main maize growing areas were assessed with regard to their suitability for coexistent production of GM and non-GM maize. The regions were: Oberbayern (DE), Ortenau (DE), Weser-Ems region (DE), Alsace (FR), Brittany (FR), upper Normandie (FR), Devon (UK), Dorset (UK), Somerset (UK), Burgas district (BG), Dobrich district (BG) and Varna district (BG). In each region, a connected area of 900 ha was evaluated. The analyzed zones of Devon and the Dobrich and Varna districts were found to be suitable, while those in Alsace, the Ortenau and the Burgas district were assessed to be suitable to only a limited extent for coexistent maize cultivation. Areas in the other regions were evaluated as partially suitable but there is a need for further coexistence or confinement measures to reduce cross-pollination.

Introduction

For coexistence between conventional, organic and genetically modified (GM) maize cultivation, the spread of GM maize as a result of pollen dispersal should be restricted. The rate of pollen-mediated gene flow between adjacent GM and non-GM maize fields depends on, beside other factors, isolation distance, size of fields and structure of field boundaries. Therefore, a survey of fields in main maize growing areas (regions with the most maize acreage in a country) in four European countries (Germany, France, the United Kingdom and Bulgaria) was conducted. The regions were selected on the basis of the EU`s statistics and the availability of well-evaluable images. The aim of the study was to assess maize growing regions with regard to their suitability for coexistence of different maize varieties, including GM und non-GM maize. We obtained the results by analyzing areas of 900 hectares. To assess the regions as a whole, further studies would be necessary.

Methods

Firstly, we carried out research on the basis of the EU statistics (European Commission of the European Union 2009) to find the main maize growing regions in the four countries. Orthophotos from the land survey agencies or pictures from Google Earth were used for the study (only pictures taken from July-August were evaluable, because in these months the maize rows can be recognized). We analyzed a connected area of 900 ha (selected randomly) in: Oberbayern (DE), Ortenau (DE), Weser-Ems region (DE), Alsace (FR), Brittany (FR), upper Normandie (FR), Devon (UK), Dorset (UK), Somerset (UK), Burgas district (BG), Dobrich district (BG) and Varna district (BG) (all images dated from 2006, only the Bulgarian photos were taken 2004). Altogether we counted 865 fields and 3045 ha under maize cultivation.

Results and discussion

Isolated fields, i.e. fields with a distance of at least 20 m and 150 m to the next maize field, were counted. The first value was chosen because it is generally assumed that a distance of 20–50 m from a pollen source is sufficient to comply with the labeling threshold of 0.9 % in maize (Devos et al. 2005; Hüsken et al. 2007). The second value is the mandatory minimum distance in Germany for cultivation of GM crops (Gentechnik-Pflanzenerzeugungsverordung 2008). The results showed that most isolated fields (in relation to all fields in 900 ha) with a distance of > 20 m were located in Devon (35 %), the upper Normandie (33 %) and Brittany (18 %). The fewest were found in the Burgas district (0 %), Alsace (3 %), and the Weser-Ems region (5 %). The highest rates of fields with isolation distances of more than 150 m were assessed in Devon (21 %), upper Normandy (13 %) and Oberbayern (7 %). In Brittany, Ortenau and in the Burgas and Dobrich districts no isolated fields with a distance of > 150 m were found (all results in this chapter refer only to the analyzed area of the region). Amongst the 865 fields assessed only 88 (10 %) were > 20 m and 22 (3 %) > 150 m isolated from neighbouring maize fields.

Regarding the size of the fields, small recipient plots are much more prone to cross-fertilization compared to large fields with a dense pollen cloud acting as a physical barrier and competitor for incoming pollen (Devos et al. 2005). As a result, large fields are favorable for coexistent maize production. The largest fields could be found in the Bulgarian regions near Dobrich and Varna (Figure 1). In the Weser-Ems region, in Dorset and Devon the median value of maize fields was 4.5–5 ha. Many small fields were assessed in Ortenau, Alsace and Somerset (median value ~ 2 ha).

Higher vegetation on field boundaries forms a physical pollen barrier and could reduce the mandatory minimum isolation distance (Devos et al. 2005). Jones & Brooks (1952) and Arrit et al. (2007) reported the effectiveness of higher vegetation to reduce cross pollination. In the present study, only the boundaries with other maize fields were evaluated (we found 1381 maize-to-maize boundaries out of 865 fields). Beside the type of boundary the distance to the next maize field was measured.

A high percentage of the boundaries were grass strips alone. Although there exists no pollen barrier, a grass strip around a pollen source may reduce the cross pollination considerably (Ludy & Lang 2004). Grass strips were common in the Burgas district, Alsace and Ortenau (Figure 2) and unusual in the three UK and the other two Bulgarian regions.

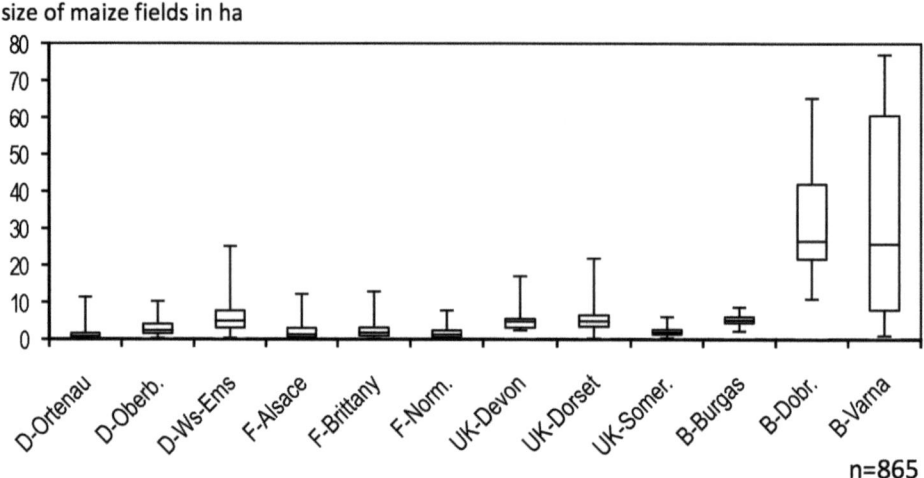

Fig. 1: Field sizes in different maize growing areas (each 900 ha), shown with boxplots.

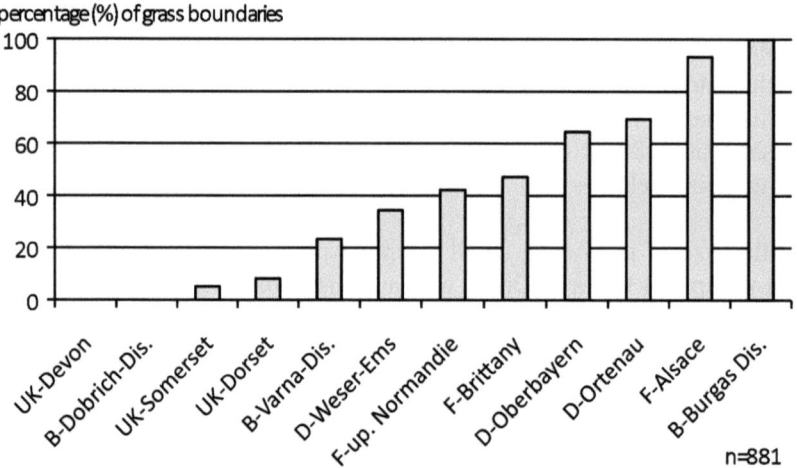

Fig. 2: Percentage of grass strip boundaries between adjacent maize fields; distance to the next maize field: 0.5–10 m.

Another option is a hedge growing on the boundary, preventing pollen flow up to a certain height. Hedges were common in the three English regions (the management of hedges around fields is supported by the government) and in the Dobrich district (Figure 3). In addition, on every third boundary in the Weser-Ems region grew a hedge. Hedges were absent in Alsace, upper Normandie and the Burgas district.

On many boundaries around maize fields one or more rows of trees have been found which can reduce the pollen flow by more than 50 % (Jones & Brooks 1952). One or more rows of trees (distance to the next maize field: < 20 m) around a maize field were common in the Dobrich and Varna districts, in Devon and also in the Weser-Ems region (Figure 4). One row of trees (distance to the next maize field: < 10 m) was frequent in Somerset, often combined with a path along the maize field. In the Burgas district, upper Normandie and Alsace rows of trees around maize fields were unusual.

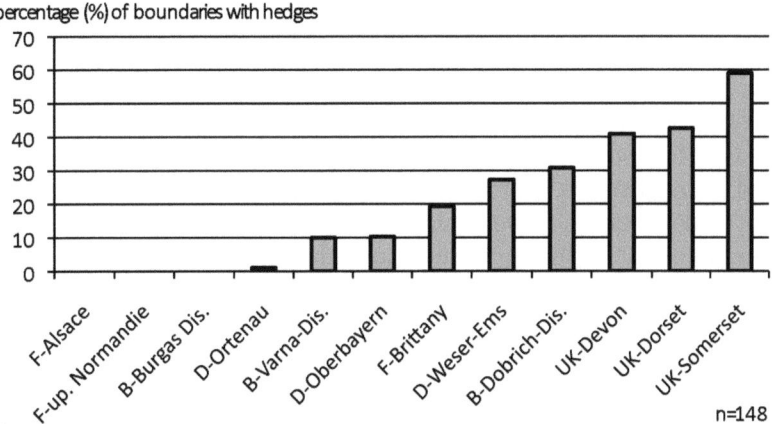

Fig. 3: Percentage of field boundaries with hedges between adjacent maize fields; distance to the next maize field: 2–10 m.

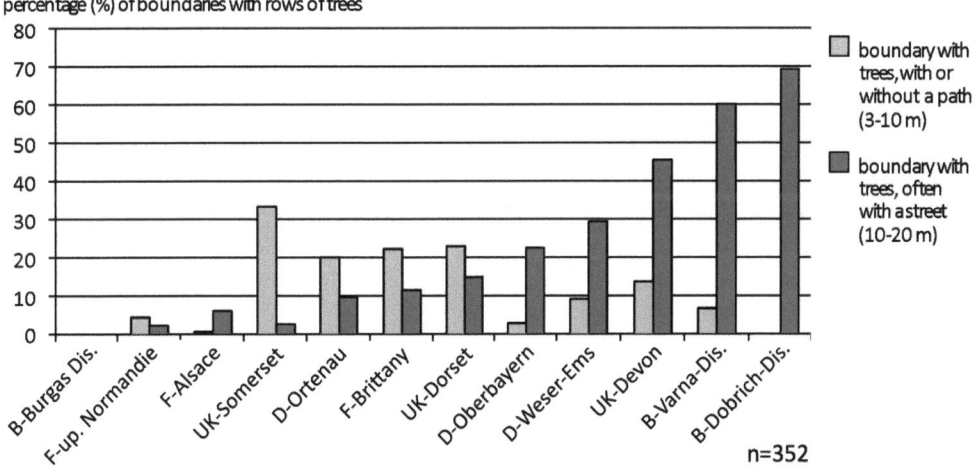

Fig. 4: Percentage of field boundaries with rows of trees between adjacent maize fields.

Conclusion

This study aims in the assessment of selected maize growing regions in Europe with regard to their suitability for coexistence of different maize varieties, using the example of the evaluated twelve 900 ha areas. In addition to isolation distance and the nature of boundaries between neighbouring maize fields several other features affect the rate of cross-fertilization in maize, for example the shape of the fields, the alignment of donor-

and recipient fields and the prevailing wind direction (Devos et al. 2005). According to the methodology of this study, these properties could not be considered.

Suitable to only a limited extend for coexistent maize production were the analyzed areas in Alsace, the Ortenau and the Burgas district, because the properties of the fields (mainly small, not isolated, short vegetation on boundaries to adjacent maize fields) promote outcrossing. Partially suitable were the areas in Oberbayern, the Weser-Ems region, Brittany, upper Normandie, Dorset and Somerset. They had features which reduce (for example higher vegetation on boundaries) but also promote (often small fields) outcrossing. In both cases coexistent cultivation of maize needs further methods to minimize outcrossing (for example increasing of separation distance or cultivation of CMS-hybrids). More suitable for coexistence were the evaluated areas in Devon and the Dobrich and Varna districts. The positive features (mainly large fields, gaps to the next maize field predominantly > 10 m, high vegetation on boundaries to adjacent maize fields) make a coexistent maize production with a low rate of cross-pollination conceivable.

Acknowledgement

Our studies were conducted within the project "Co-Extra – GM and non-GM supply chains, their Coexistence and Traceability", supported by the EU (proj. no. 007158).

References

Arrit R.W., Astini J., Clark C.A., Westgabe J.M.E., Goggi A.S. (2007) Biological windbreaks for pollen confinement. Third International Conference of Coexistence between Genetically Modified (GM) and non-GM based agricultural supply chains (Ed.: European Commission), Conference Paper, Sevilla, Spain, 131–134.

Devos Y., Reheul D., de Schrijver A. (2005) The co-existence between transgenic and non-transgenic maize in the European Union: a focus on pollen flow and cross-fertilization. Environmental Biosafety Research 4: 71–87.

European Commission of the European Union (2009): Eurostat. Landwirtschaftliche Fläche: Anzahl der Betriebe und Flächen nach Flächennutzung. http://epp.eurostat.ec.europa.

Gentechnik-Pflanzenerzeugungsverordnung (2008) Verordnung über die gute fachliche Praxis bei der Erzeugung gentechnisch veränderter Pflanzen (GenTPflEV). Bundesministerium für Justiz (ed.), 7.4.2008.

Jones M., Brooks J. (1952) Effect of tree barriers on outcrossing in corn. Oklahoma Agr. Exp. Sta. Tech. Bul. 45.

Hüsken A., Ammann K., Messeguer J., Papa R., Robson P., Schiemann J., Squire G., Stamp P., Sweet J., Wilhelm R. (2007) A major European synthesis of data on pollen and seed mediated gene flow in maize in the SIGMEA project. In: Stein A.J., Rodríguez-Cerezo E. (Eds.) Book of abstracts of the third International Conference on Coexistence between Genetically Modified (GM) and non-GM-based Agricultural Supply Chains, European Commission, pp. 53–56.

Ludy C., Lang A. (2004) Dispersion and deposition of Bt maize pollen in field margins. Journal of Plant Disease and Protection 11(5): 417–428.

Breckling, B. & Verhoeven, R. (2010) Implications of GM-Crop Cultivation at Large Spatial Scales. Theorie in der Ökologie 16. Frankfurt, Peter Lang.

Is there any room for alternatives? Socio-economic implications of GMOs cultivation at large-scale – Case study in Spain[1]

Rosa Binimelis[a], Iliana Monterroso[a,b] & Mariel Vilella[c]
([a]Institute for Ecological Economics and Political Ecology; Barcelona, Spain; [b]Latin American Faculty of Social Sciences FLACSO-Guatemala; [c]Pompeu Fabra University, Barcelona, Spain. – rosa.binimelis@gmx.net)

Abstract

The authorisation procedure for the placing on the market of GMOs in Europe has come under strong criticism. One of the major critiques is that socio-economics implications of GMOs cultivation are not taken into account in the decision-making process. For this reason, in December 2008, the EU Environmental Council (2008) invited the Member States to "collect and exchange relevant information on socio-economic implications of the placing on the market of GMOs including socio-economic benefits and risks and agronomic sustainability", while requested the European Commission to submit a report based on these contributions. The inclusion of socio-economic aspects has also been urged by several Member States in the Environment and Agriculture Councils.

In this context, the objective of this paper is contribute to this discussion by analysing the socio-economic implications of GMOs cultivation in Catalonia and Aragon (NE of Spain), GM maize has been commercially grown for the last 12 years. In 2009, the number of hectares grown with GM maize was 28.240 and 29.540 respectively. In the European context, this implies a unique experience of large-scale GMOs cultivation.

For doing so, the paper focuses on the implementation of the coexistence policy framework. The concept of coexistence was introduced by the European Commission as a compromise policy that, through the establishment of science-based technical measures, should ensure that farmers can freely choose between different production systems (organic, conventional or using GMO crops). However, this concept is far from bringing along a consensus and has become, itself, a major battlefield at the policy arena (Levidow & Boschert 2007).

In despite the wide socio-economic implications of GMOs cultivation side-by-side to other forms of agriculture, coexistence has been framed as a technical and "objective" issue. This is done by means of individualising the choice, reducing the risks to those economic aspects that can be individualised and quantified (Binimelis 2008), and by

---

1 Extended abstract: A full paper is submitted to UWSF – Zeitschrift für Umweltchemie und Ökotoxikologie, Series: Implications of GMO-cultivation and monitoring. Springer-Verlag.

excluding other rationales and criteria using "science" as a source of legitimation (Devos et al. 2008), restricting the participation in the decision-making to those actors with an "economic" stake (Lee 2008). It supposes that coexistence measures are independent of socio-cultural, political and economic conditions, assuming this independency as a source for legitimacy.

However, this paper demonstrates the lack of a common purpose (e.g. the definition of what agricultural model we are targeting as a society) inherent to the different approaches for coexistence. Any formulation of rules to ensure coexistence depends on the options taken at the policy level. In a context of high social unrest, the coexistence policy reinforces the dominant productivist agricultural model over those options based on alternative rationales.

References

Binimelis R. (2008) Coexistence of Plants and Coexistence of Farmers: Is an Individual Choice Possible? Journal of Agricultural and Environmental Ethics 21(5): 437–457.
Devos Y., Maeseele P., Reheul D., van Speybroeck L., de Waele D. (2008) Ethics in the societal debate on Genetically Modified Organisms: A (re)quest for sense and sensibility. Journal of Agricultural and Environmental Ethics 21: 29–61.
EU Environmental Council (2008) Council Conclusions on Genetically Modified Organisms (GMOs). www.consilium.europa.eu/ueDocs/cms_Data/docs/pressData/en/envir/104509.pdf.
Lee M. (2008) The Governance of Coexistence Between GMOS and Other Forms of Agriculture: A Purely Economic Issue? Journal of Environmental Law 20(2): 193–212.
Levidow L., Boschert K. (2008) Coexistence or contradiction? GM crops versus alternative agricultures in Europe. Geoforum 39(1): 174–190.

Development of the indicator "Genetic engineering in agriculture"

Christiane Eschenbach & Wilhelm Windhorst
(Ecology Centre, University of Kiel, Germany. – ceschenbach@ecology.uni-kiel.de)

Introduction

In order to allow the assessment of trends in GMO-cropping the indicator "Genetic engineering in agriculture" has been developed to mirror the spatial extension of GMO cropping in Germany. The indicator seeks to operationalise the precautionary principle and the main objective is to unveil potential threats on biodiversity. The development of the indicator was initiated by the Federal Agency for Nature Conservation (BFN) in accordance with the German National Biodiversity Strategy.

Methods and data base

In order to base the indication of GMO cropping on a broad base and to account for a wide range of ecosystemic and spatial interactions, ecology and cropping of Bt-maize, herbicide tolerant rape and potatoes with altered starch composition were analysed. The development of the indicator is related to commercial cultivation (not experimental releases) and focused on agriculture, but forestry can be integrated. As genetic modifications of other species have to be expected in future the indicator construction is open concerning GM-treats. Hence, newly approved GMO by Germany and the EU can become part of the indicator calculation as soon as regular farming will take place – without changing the character of the indicator.

The core data base for the development of the spatial indicator is the site register for GMO cropping provided by the Federal Agency for Consumer protection and Food safety (BVL). The site register was established in 2005 and will be maintained on a long term basis and provide site specific data for each field cropped with GMO in Germany. However, as precise geo-referencing of the GMO fields is not supported by the site register yet – and further cadastral information was not made available – the indicator was developed without geo-referenced spatial data. Should further land survey data become available in future, the devised calculations for the indicator could be made with higher accuracy and would allow evaluations of spatial ecological interactions without jeopardising the envisaged long term consistency of the data sets.

Results

The need for this indicator is recognised in the German National Biodiversity Strategy with the sub-indicators (1) "Land cultivated with genetic engineering" and (2) "Agricultural land without the cultivation of genetically modified plants". The performed literature review (e.g. Züghart & Breckling 2003; Devos et al. 2004; Messéan et al. 2006) and the analysis of recent R&D activities in GMO risk analysis revealed that spatial interactions of GMO with adjacent ecosystems are to be expected, but may decline with distance. To account for these exposed areas in the vicinity of GMO-cropping, which are of special relevance for the objects of legal protection in the German Federal Nature Conservation Act (BNatSchG 2002), a third sub-indicator (3) "GMO-exposed area" has been devised. In order to determine the "GMO-exposed area" a literature study was performed to identify those direct interactions of GM-maize, GM-rape, and GM-potatoes with their neighbourhoods representing a pressure with the potential to impact the biodiversity. The analysis considered scientific evidence, level of local relatedness and data availability at present and in future.

The size of the "GMO-exposed area" is determined by the ways how GMO and their residues spatially interact with their environment. These interactions are described as pressures to the exposed area which may, but must not necessarily impact the exposed biodiversity. The following interactions have been analysed and classified as suitable for the calculation of indicator values, only suitable for selected species or not suitable: Seed impurity, wind induced pollen transfer to neighbouring fields, pollen transfer by honey bees to adjacent areas, long distant transport of pollen by wind, transport losses of seed and/or harvest, transport of plant residues by birds and other animals, transport of GMO residues via water and wind, hybridisation potential, seed bank development, volunteers in follow up crops, possible establishment of feral GMO populations, possible hybridisation with non GMO cultivares and/or native species, accumulation of transgenes or GMO residues (for example Bt-toxin) in soils. Hence, the sub-indicator "GMO-exposed area" comprises the GMO-cropped area plus a buffer area defined by the spatial extent and intensity of interactions between GMO and adjacent biodiversity.

Further assumptions had to be made to allow replicable calculations. As no geo-referenced data on the field shape could be provided, the calculation of the sub-indicator "GMO-exposed area" is based on the assumption of quadratic fields. With a growing share of GMO-cropping there is an increasing probability of spatial overlap of buffer zones and due to crop rotation and to time lags in decomposition of plant residues or due to accumulations in the seed bank also time dependent overlaps have to be considered. These phenomena are accounted for by assuming a linear increment and by a single event analysis, respectively.

Additionally, criteria were assembled and developed to decide on the distances to calculate the size of the buffer zone surrounding GMO cropped fields. For each of the above interactions classified as suitable a procedure has to be chosen to set this distance. Scenarios for different crops based on highly resolved and geo-referenced data sets for the federal state of Schleswig-Holstein (15,500 km^2) have been elaborated to underpin this decision process. Accounting for different principle strategies to yield a replicable decision (technical detection limit, minimum distances acc. to existing law regulations, chart analysis, educated guess, impact analysis), the BfN eventually decided to set the distance to calculate the Bt-maize exposed area at 1,000 metres. This distance should include all different interactions considered above. The determination of distances to be used for further GMO will be based on decisions by expert groups.

The sub-indicator "Land cultivated with genetic engineering" (1) is calculated on the data on GMO cropped fields, provided by the site register and represents the area of all GMO-cropped fields as reported by the farmers. The former sub-indicator "Agricultural land without the cultivation of genetically modified plants (2)" has been defined as the area where no GMO or transgenic residues exist or can be neglected due to irrelevance: "Non-GMO exposed area".

For all three sub-indicators the relative share of the German territory is calculated and presented as "area in %" on an annual basis. The calculated values for 2008 are: (1) "Land cultivated with genetic engineering": 0.0089 %, (2) GMO-exposed area: 0.3013 %, (3) Non-GMO exposed area: 99.699 %.

Conclusions

Including the sub-indicator "GMO-exposed area" the indicator "Genetic engineering in agriculture" accounts for the objectives of legal protection in the German Federal Nature Conservation Act (BNatSchG 2002) and fulfills the purpose of environmental indicators to maintain, restore or improve specified qualities of ecosystems (Müller & Wiggering 2004).

References

BNatSchG (2002) German Federal Nature Conservation Act (Gesetz zur Neuordnung des Rechts des Naturschutzes und der Landschaftspflege und zur Anpassung anderer Rechtsvorschriften in der Fassung der Bekanntmachung vom 25. März 2002.).

Devos Y., Reheul D., De Schrijver A., Cors F., Moens W. (2004) Management of herbicide-tolerant oilseed rape in Europe: a case study on minimizing vertical gene flow. Environ. Biosafety Res. 3: 135–148.

Messéan A., Angevin F., Gomez-Barbero M., Menrad K., Rodriguez-Cerezo E. (2006) New case studies on the coexistence of GM and non-GM crops in European agriculture. European Commission, Joint Research Centre.

Müller F., Wiggering H. (2004) Umweltziele und Indikatoren. Begriffe, Methoden, Aufgaben und Probleme. In: Wiggering H., Müller F. (ed) Umweltziele und Indikatoren. Wissenschaftliche Anforderungen an ihre Festlegung und Fallbeispiele. Berlin, Heidelberg, Springer, 648 pp.

Züghart W., Breckling B. (2003) Konzeptionelle Entwicklung eines Monitoring von Umweltwirkungen transgener Kulturpflanzen. UBA-Texte 50/03.

New pest in crop caused by large scale cultivation of Bt corn

Christoph Then
(Testbiotech, München, Germany. – christoph.then@testbiotech.org)

Abstract

Since the year 2000 it has been observed in the United States that genetically engineered corn (maize) plants expressing the Bt toxin classified as Cry1Ab are being infested by the larvae of the western bean cutworm (*Striacosta albicosta*). Originally this pest only occurred within narrowly confined regions and caused no major problems in corn. For several years now, however, this pest has been spreading into more and more regions within the North American Corn Belt and causing substantial economic damage. There are empirical findings that the new pest has been caused by large scale cultivation of genetically engineered plants expressing Cry1Ab. This is considered to be a specific case of 'pest replacement'. In this case the corn earwom (*Helicoverpa zea*), a naturally occurring competitor of the western bean cutworm has been accidentally eliminated by the extensive cultivation of Bt corn. Under these circumstances the new pest is able to spread on a large scale and to infest crops heavily. This contribution reviews facts and findings and discusses possible strategies to counteract this new pest and serves as an overview on recent pest management problems in Bt crops.

New pest spreads through US corn belt

Since the year 2000 it has been observed that genetically engineered corn expressing the Bt toxin Cry1Ab is being infested by western bean cutworm (*Striacosta albicosta*) (Rice 2000; O'Rourke & Hutchison 2000). The western bean cutworm was historically only found in some regions and caused only minor problems in corn. At present, it is spreading into more and more states of the United States where it is causing significant economic damage. In 2006, a scientific publication reported extensive damage in South Dakota (Catangui & Berg 2006). In the meantime, western bean cutworm damage has been documented for almost all states in the North American Corn Belt. States affected for example include Iowa, Missouri, Minnesota, Wisconsin, Indiana, Michigan and Ohio (Eichenseer et al. 2008).

Pest replacement in genetically engineered corn

There are several studies explaining how the spread of the western bean cutworm is fostered by growing genetically engineered corn. Apparently it is a case of so called pest replacement (Butzen et al. 2007). This is a phenomenon previously observed in intensive agriculture, where there is a massive use of pesticides. Pest replacement opens up new ecological niches in which other competitors (pests) can thrive. In this case Cry1Ab expressed by genetically engineered corn (YieldGard © Monsanto) is not only active against the European corn borer but also active against the corn earworm (*Helicoverpa zea*). This latter pest feeds not only on corn but is also cannibalistic to other pest insects such as the western bean cutworm (Rice & Dorhout 2006). The corn earworm is sensitive to the Bt toxin Cry1Ab, while the western bean cutworm is not. Thus the equilibrium situation between the two insect pests can be significantly changed. Interaction between the western bean cutworm and the corn earworm was confirmed in 2010 (Dorhout & Rice 2010), showing the spread of the western bean cutworm is in fact fostered by the cultivation of Bt corn expressing Cry1Ab. Damages caused by the western bean cutworm can even exceed those caused by the European corn borer in conventional plants (Catangui & Berg 2006). There are other reports of increasing problems with Bt crops which have been grown permanently on a large scale. For example in 2006, shifts in pest insects were reported in Bt cotton grown in China (Lu et al., 2010). Tabashnik et al. (2009) present several cases of resistance of pest insects to Bt crops in the fields.

Tab. 1: Some recent publications about pest management problems in Bt crops.

Source (Year)	Species	Crop/ Region	Effect
O'Rourke & Hutchison (2000)	Western bean cutworm	Corn / USA (Minnesota)	Pest replacement
Dorhaut & Rice (2004)	Western bean cutworm	Corn / USA (Illinois, Missouri)	Pest replacement
Catangui & Berg (2006)	Western bean cutworm	Corn / USA (South Dakota)	Pest replacement
Li et al (2007)	Cotton bollworm	Cotton/ China	Higher tolerance (Cry1Ac)
Wang et al (2008)	Mirid bug	Cotton / China	Secondary pests
Di Fonzo & Hammond, (2008)	Western bean cutworm	Corn / USA (Michigan, Ohio)	Pest replacement
Tabashnik et al (2009)	Fall armyworm	Corn / Puerto Rico	Resistance (Cry1F)
Tabashnik et al (2009)	Maize stalk borer	Corn/ South Africa	Resistance (Cry1Ab)
Tabashnik et al (2009)	Cotton bollworm	Cotton/ USA	Resistance (Cry1Ac, Cry2Ab)
Zhao et al, (2010)	Aphids, spider mites, lygus bugs	Cotton/ China	Secondary pests
Lu et al, 2010	Mirid bug	Cotton/ China	Secondary pests
Monsanto (2010)	Pink bollworm	Cotton/ India	Resistance (Cry1Ac)

Points for discussion: Industry's solution

Pioneer Hi-Bred and Dow AgroSciences are marketing a further corn hybrid in the USA, so-called 'Herculex' Corn, which expresses another variant of the Bt toxin (Cry1F), meant to be effective against western bean cutworm larvae. This genetically engineered corn has been grown commercially in the US since 2001. But western bean cutworm infestation can be curbed by growing the new corn hybrid, not completely prevented, since Herculex is only 80 to 90 percent effective against western bean cutworm (Eichenseer et al. 2008). Additionally in 2009, the USA and Canada licensed a genetically engineered corn hybrid with eight gene constructs incorporating six different Bt toxins called 'SmartStax'. In relation to the control of the western bean cutworm the active ingredient in SmartStax is also Cry1F – thus it inherits the same deficiency as the 'Herculex' plants. In addition, those plants also produce Cry1Ab thereby suppressing the natural competitor of the western bean cutworm. Large scale cultivation of crops like Herculex or SmartStax can cause the less sensitive larvae of the western bean cutworm being selected systematically and spreading rapidly throughout the population. Thus the so called solutions could even aggravate the current situation. The spread of the western bean cutworm will mean good business for companies even if growing genetically engineered corn can not deliver as promised. The company Dupont, for instance, which owns the seed producer Pioneer, is advertising not only genetically engineered corn but also insecticides such as 'Asana XL' to control the new pest.

Some more points for discussion

There is increasing evidence that strategies used for large scale cultivation of Bt plants such as corn, cotton or rice need to be reassessed. In a publication in the magazine Nature (Qiu 2008) plans for Bt rice cultivation in China are questioned because many of the already known pest insects could not be controlled by the Bt produced in the plants. In the article, a researcher from the International Rice Research Institute IRRI, the Philippines, raised a very basic question concerning the general strategy of growing Bt-plants: "Pests thrive where biodiversity is at peril. Instead of genetic engineering, why don't we engineer the ecology by increasing biodiversity?" There is a growing need to find alternatives to current practises. alternatives to current practices. There might be several reasons for pest replacement, like climate change, change in agricultural practices, but the large scale growing of Bt crops seems to play a major role here. Pest replacement and pest resistance seem to be an inevitable consequence of any strategy that continuously tries to suppress or eliminate pest organisms. This is especially true for the strategy underlying the usage of Bt crops or plants expressing VIP toxins (as propagated by company of Syngenta), since the release of the toxin is not targeted and time limited, but implies permanent exposure throughout the whole period of cultivation. The ecosystem can be destabilised by suppressing certain insects at the same time the door is opened to pest replacement and pest resistance in major pest insects. Subsequently, farmers will end up doing two things – buying expensive seeds to grow multi-stacked Bt-plants and spraying hazardous pesticides.

References

Butzen S., Dorhout D., Davis P. (2007) Spread of Western Bean Cutworm in the U.S. Corn Belt. Crop Insights Vol. 17 No. 10.
   http://www.mccormickcompany.net/pioneer/cropinsights/63.pdf.

Catangui M.A., Berg R.K. (2006) Western bean cutworm, Striacosta albicosta (Smith) (Lepidoptera: Noctuidae), as a potential pest of transgenic Cry1Ab Bacillus thuringiensis corn hybrids in South Dakota. Environmental Entomology 35: 1439–1452.

Di Fonzo C.D., Hammond R. (2008) Range expansion of western bean cutworm, Striacosta albicosta (Noctuidae), into Michigan and Ohio. Crop Mgt. Online: doi: 10.1094/CM-2008-0519-01-B.

Dorhout D.L., Rice M.E. (2004) First report of western bean cutworm, Richia albicosta (Noctuidae) in Illinois and Missouri. Crop Management.
http://www.plantmanagementnetwork.org/pub/cm/brief/2004/cutworm.

Dorhout D.L., Rice M.E. (2010) Intraguild Competition and Enhanced Survival of Western Bean Cutworm (Lepidoptera: Noctuidae) on Transgenic Cry1Ab (MON810) Bacillus thuringiensis Corn. Journal of Economic Entomology 103: 54–62. doi: 10.1603/EC09247.

Eichenseer H., Strohbehn R., Burks J. (2008) Frequency and Severity of Western Bean Cutworm (Lepidoptera: Noctuidae) Ear Damage in Transgenic Corn Hybrids Expressing Different Bacillus thuringiensis Cry Toxins. Journal of Economic Entomology 101 (2): 555–563.

Lu Y., Wu K., Jiang Y., Xia B., Li P., Feng H., Wyckhuys K.A.G., Guo Y. (2010) Mirid Bug Outbreaks in Multiple Crops Correlated with Wide-Scale Adoption of Bt Cotton in China. Science 328 (5982): 1151–1154.

Li G., Wu K., Gould F., Wang J., Miao J., Gao X., Guo Y. (2007) Increasing tolerance to Cry1Ac cotton from cotton bollworm Helicoverpa armigera, was confirmed in Bt cotton farming area of China. Ecological Entomology 32: 366–375.

Monsanto (2010) Cotton in India.
http://www.monsanto.com/monsanto_today/for_the_record/india_pink_bollworm.asp.

O'Rourke P.K., Hutchison W.D. (2000) First report of the western bean cutworm, Richia albicosta (Smith) (Lepidoptera: Noctuidae), in Minnesota corn. J. Agric. Urban. Entomol. 17: 213–217.

Qiu J. (2008) Is China ready for GM rice? Nature 455: 850–852.

Rice M.E. (2000) Western bean cutworm hits northwest Iowa. Integrated Crop Manage. IC-484, 22: 163. Iowa State University Extension, Ames, IA.

Rice M.E., Dorhout D.L. (2006) Western bean cutworm in Iowa, Illinois, Indiana and now Ohio: Did biotech corn influence the spread of this pest? In 2006 Integrated Crop Management Conference. Iowa State University: 165–172.

Tabashnik B.E., Van Rensburg J.B.J., Carrière Y. (2009) Field-Evolved Insect Resistance to Bt Crops: Definition, Theory, and Data. J. Econ. Entomol. 102(6): 2011–2025.

Wang S., Just D.R., Pinstrup-Andersen P., (2008) Bt-Cotton and secondary pests. J. Biotechnol. 10: 113–120.

Zhao J.H., Ho P., Azadi H. (2010) Benefits of Bt cotton counterbalanced by secondary pests? Perceptions of ecological change in China, Environ. Monit. Assess. DOI 10.1007/s10661-010-1439-y.

Chapter IV

Setting the frames: integrative interdisciplinary approaches

Breckling, B. & Verhoeven, R. (2010) Implications of GM-Crop Cultivation at Large Spatial Scales. Theorie in der Ökologie 16. Frankfurt, Peter Lang.

The Danish coexistence regulation and the Danish farmers attitude towards GMO

Morten Gylling
(Institute of Food and Resource Economics, University of Copenhagen, Denmark. – gylling@foi.dk)

Introduction

As a consequence of the revised release directive (Directive 2001/18/EEC) published in April 2001, The Danish Parliament in May 2002 passed the *Act on the amendment of the Act on environment and gene technology*, the Act implements the new release directive. Into the Act was also inserted a provision that the Minister for Food, Agriculture and Fisheries lays down regulations that within the framework of EU legislation severely restricts the risk of dispersal to other fields, including organic fields.

The preparation of a Danish coexistence strategy

In June 2002, the Danish Minister for Food, Agriculture and Fisheries took the initiative to prepare a strategy of coexistence of genetically modified, conventional and organic crops. In this context Coexistence was defined as *the precautions necessary to give the possibility to ensure simultaneous production in a region of GM, conventional non-GM and organic crops and thus maintaining the consumers and farmers free choice between GM and non-GM.*

Purpose of the co-existence strategy work
The aim of the strategy work was to describe possibilities and conditions for a commercial use of the gene technology in agriculture that supports the free choice of consumers and the potential of current production systems. Further, the work was to establish a basis for decisions that may constitute the starting point of regulation. Finally, it was the intention that the strategy were prepared in a continuous dialogue with the public. For this work, a Working Group, a Strategy Group and a Contact Group were appointed, it was the intention that the strategy was prepared in a continuous open dialogue with all stakeholders.

Organisation of the work
The Working Group was set up, consisting of the Danish Institute of Agricultural Sciences, the Danish Plant Directorate, the Danish Research Institute of Food Economics, National Environmental Research Institute, Denmark, and scientists from the Royal Veterinary and Agricultural University and Risø National Laboratory.

The aim of the Working Group was to ensure an adequate scientific analysis of the dispersal problems and possible control measures that take Danish conditions as their starting points.

The assignment of the Working Group was to:
- Perform a scientific evaluation of sources of dispersal from genetically modified productions to conventional and organic productions.
- Evaluate the extent of dispersal and the need of control measures.
- Identify and evaluate possible control measures to ensure co-existence of genetically modified, conventional and organic production systems.

The Working Group should involve the Contact Group in accordance with the task specification for this Group. The evaluation were to be carried out on the basis of Danish conditions and on the basis of the knowledge that was available regarding this subject and the consequences for business economics were to be evaluated.

The Working Group decided that the evaluation should be restricted to:
- Danish plant production of significant agricultural crops and seed growing of selected vegetables, but no fruit, berries or forest trees.
- Calculations regarding business economics comprising primary production, i.e. multiplication up to and including vegetable production (first stage of distribution). In four selected cases (sugar, rapeseed oil, feed wheat, and an actual food product), however, calculations of costs further on in the production chain were made.

The Group did not make any recommendation on who should cover extra costs in connection with a possible adventitious admixture of GM or who should cover any costs that may be incurred in connection with monitoring and control. Neither does it make any recommendations on where costs should be placed in connection with separation distances, buffer zones, etc.

The continued evaluation work
The Working Group presented the 1st edition of its report of January 9, 2003 at an Expert Hearing arranged by the Ministry of Food, Agriculture and Fisheries at Christiansborg Palace on January 21, 2003. Based on the conclusions in the first version of the report of the Working Group and from the Expert Hearing, the Minister for Food, Agriculture and Fisheries decided that the work on the evaluation should continue with a view to elaborate and update the report, which were performed during 2003. The final report presents a number of crop specific coexistence measures. The recommendations are as part based on the regulations for production of certified seeds and a threshold of adventitious GM presence in seeds of 0.3 % (cross pollinating), 0.5 % (self pollinating) and 0.7 % (field peas).

Legal framework and liability issues for Danish GM farmers

The Danish coexistence act was passed through Parliament in June 2004 with support from a broad political majority (Act no. 436 of June 9, 2004).

The implementation of the Act is regulated by statutory order No. 220 of March 31, 2005 (growing of genetically modified crops). A statutory order No. 1170 of December 9, 2005 regulates the compensation scheme. The coexistence act and the statutory orders create a legal regulatory framework for the Danish GM farmers regarding cultivation and liability. The Coexistence act offers legal and economic protection to both the GM-growers and the non-GM growers.

The coexistence act contains a number of elements to full fill the national strategy:
- mandatory education of GM-growers (GM-drivers licence),
- authorisation of GM-growers,
- registration of GM-growers and GM fields (public register),
- good agricultural practices (see Tolstrup et.al. 2003),
- crop specific separation distances between GM and non GM-crops,
- a compensation scheme for farmers suffering losses from adventitious presence of GM-crops into non GM-crops from neighbouring GM-crops,
- a fee of 100 DKK (13.4 €) per hectare GM-crops grown to finance the compensation scheme.

The compensation scheme

The compensation scheme is administrated by The Danish Plant Directorate, and is financed by a fund in to which the GM farmer has to pay 100 DKK per hectare GM crop cultivated. It is a future goal that the compensation scheme will be replaced by some kind of commercial insurance model, however, at present the insurance companies are not interested in marketing this kind of insurance.

GM-growers fulfilling the rules laid down in the act will on the one hand not be liable for adventitious presence of GM in neighbouring crops, on the other hand should adventitious presence of GMO in neighbouring non GM-crops occur the non GM-grower will get compensation from the compensation scheme, without having to go through to court system. Compensation from the compensation fund will be paid for economic losses caused by adventitious presence of GM above 0.9 % in conventional and organic crops within 1.5 times the crop specific separation distance. The compensation will be paid from the compensation fund regardless of whether the GM grower has complied with the coexistence regulation or not. However, in the case of non compliance with the regulation the GM grower will be held liable and it is up to the Danish Plant Directorate to decide which legal action to take.

To apply for compensation a non GM grower must supply samples and analysis of the crop, the cost for this will be compensated if the non GM grower is found entitled for compensation.

The Coexistence Act and the Compensation Scheme covers the primary crop production up to farm gate. After the farm gate the general Food and Feed regulations apply by which the "one step back and one step forward" principle should secure full traceability in the supply/processing chain.

Danish farmers attitudes towards GMO

The act was well received by the agricultural sector and is seen as a regulatory framework that gives the opportunity to grow GM-crops and at the same time protects GM-growers as well as non GM-growers.

A survey among farmers (Lawson 2009) showed that:
- 45 % are positive, 27 % are neutral and 28 % are negative towards GMO.
- The main concern is to avoid affecting other production types.
- > 50 % do not expect improved profitability (30 % do).
- A majority of farmers expect input savings from GM-crops.
- Adaptation of GM-crops seems to be linked with expected economic return.

So far no GM-crops are grown commercially in Denmark, but more than 300 farmers have at present been trained to grow GM crops and the "GM drivers licence" will be part of the future education of young farmers.

References

Danish Parliament (2004). Act no. 436 of June 9 2004.
Danish Ministry of Food, Agriculture and Fisheries (2005). Statutory order No. 220 of March 31, 2005 (growing of genetically modified crops).
Danish Ministry of Food, Agriculture and Fisheries (2005). Statutory order No. 1170 of December 9, 2005 (regulates the compensation scheme).
Lawson L.G., Larsen A.S., Pedersen S.M & Gylling, M. (2009). Perceptions of genetically modified crops among Danish farmers. Food Economics 6(2).
Tolstrup K., Andersen S.B., Boelt B., Buus M., Gylling M., Holm P.B., Kjellson G., Pedersen S., Østergård H. & Mikkelsen S.A. (2003) Report from the Danish Working group on the Co-existence of Genetically Modifies Crops with Conventional and Organic Crops. DIAS report. Plant Production 94, November 2003.

Breckling, B. & Verhoeven, R. (2010) Implications of GM-Crop Cultivation at Large Spatial Scales. Theorie in der Ökologie 16. Frankfurt, Peter Lang.

From risk assessment to in-context trajectory evaluation: GMOs and their social implications[1]

Vincenzo Pavone[a], Joanna Goven[b] & Riccardo Guarino[c]
([a]CSIC – Consejo Superior Investigaciones Científicas, Institute of Public Policies. Madrid, Spain; [b]University of Canterbury – School of Social and Political Sciences, Christchurch, New Zealand; [c]University of Palermo – Faculty of Mathematics, Physics and Natural Sciences, Department of Botanic Sciences, Palermo, Italy. – vincenzo.pavone@cchs.csic.es)

Abstract

Over the past twenty years, biotechnologies have raised enormous expectations as well as passionate political controversies, paving the way to a strong polarization in European society and to an on-going debate on how should these technologies be assessed. Mainstream approaches have been focusing on risk-assessment procedures. According to this perspective, new technologies should be assessed in terms of their potential risk of negatively affecting human health and in terms of the environmental risks, such as cross-contamination and biodiversity preservation. Yet, the large majority of risk-assessment studies on GMOs mainly focus on animal trials, trying to detect biological or medical anomalies among the animals fed with GM products. Although many of these studies have repeatedly claimed that no significant health impact could be detected, their independence and reliability has been contested not only because they have been carried out by the same multinational corporations that produce the tested GMOs but also because the original data have not been released to the academic community for the studies to be replicated. Moreover, independent studies on GMOs have raised serious doubts about health safety in a number of different occasions (Le Curieux-Belfond et al. 2008; Seralini et al. 2009; Seralini, Cellier & Spiroux de Vendomois 2007; Gasnier et al. 2009; Heinemann & Traavik 2004; Traavik & Heinemann 2007).

Independently of whether GMOs constitute a direct threat to human health and the environment, risk-assessment approaches have reduced the evaluation of GMOs merely to a question of how much risk can a society bear for the introduction of these new products in the face of their claimed benefits but there is much more to GMOs than the risk/benefit relationship suggests (Ferretti 2009). Many reasons lay behind the emergence and diffusion of risk-assessment approaches. On the one hand, these approaches support and strengthen the technological fix attitude that affects post-industrial societies. Problems that may have a number of different social, economic or political origins are

---

[1] Extended abstract: A full paper is submitted to UWSF – Zeitschrift für Umweltchemie und Ökotoxikologie, Series: Implications of GMO-cultivation and monitoring. Springer-Verlag.

framed and addressed in terms of a technological solution that allow for a quick, effective fix that does not call into question these non-technical origins. A clear example maybe retrieved in the Syngenta website, where the issue of water scarcity and water supply all over the globe is reduced to a technical question, whose solution is offered through GM crops with reduced water absorption (www.singenta.com, "Bring plant potential to life" campaign). On the other hand, these approaches positively resonate with the tendency to delegate essentially political decisions to expert committees, which effectively divert responsibility from political actors to techno-scientific networks (Jasanoff 2003). In turn, this process de facto de-politicizes a number of controversial issues, which could otherwise threaten political consensus and stability. As a consequence, the growing momentum of risk-assessment approaches has encouraged a technocratic twist in science and technology policy, which has been criticized on a number of political and sociological grounds (Weingart 1999; Funtowicz & Liberatore 2003; Nowotny 2003; Felt at al. 2007; Levidow 2009; Ferretti & Pavone 2009).

First, it has been argued that risk-assessment approaches take the technology for granted, addressing public opposition to GMOs as a problem in itself. Instead of considering public arguments against GMOs as an opportunity to reconsider the technology from a different perspective, producing a wider and more robust assessment of GMOs' implications, the public has been addressed as the problem, calling for solutions that aimed at reducing this opposition rather than at learning from it (Felt et al. 2007; Levidow 2007).

Second, risk-assessment approaches address GMOs potential impact merely in terms of their human health-related and environmental risks. However, GMOs have also an important impact not only on the existing economic, political and social arrangements but also on the developmental trajectory of the areas selected for implantation. Technology shapes society and it is shaped continuously by it, in a mutually constitutive process that has been elsewhere described as co-production (Jasanoff 2005; Ferretti & Pavone 2009). In this co-production process, science and technology and social order emerge side by side.

Third, it has been pointed out that these technologies cannot be evaluated in abstract terms, independently of the juridical, social and economic context in which they will be implemented. Local institutional rules and practices shape technology innovation and implementation and cannot be considered equal in each and every corner of the world. Power relationships, economic interests, lack of transparency, weak rule enforcement may strongly affect not also the trajectory of implementation of a technology but also the actual repercussions that GMOs are likely to produce (Goven 2006b).

Last but not least, approaches focusing on risk do not call into question the actual trajectory that a technological innovation has followed to emerge, and the visions and imaginaries that came along with it (Mcnaghten et al. 2006; Felt et al. 2007). In other words, technological products are no neutral objects. They have been produced by specific actors, in specific contexts, in order to address a specific problem, which has been framed in such a way that a given technology makes sense as a solution. As a result of the very process triggering their emergence, technologies are loaded with social and political values. Technologies materialise certain paradigms, in fact, they "re-construct" social paradigms (ideas and assumptions about functioning) into physical matter – this is what could make the utility of a technology. It has to "fit" to the social structures managing it, and resembles the material support a social setting organises to stabilise and proceed itself, which will remain completely undetected as long as the focus of technology assessment concentrates on their risk implications. Yet, a thorough analysis of the ethical, social and political load of values and principles that each technology carries in the visions and imaginaries it promotes is a fundamental step towards a social and political assessment of risk technologies in general, and GMOs in particular.

This paper will be developed in three steps. In the first step, the limits and implications of risk-assessment approaches will be outlined and discussed. In the second step, recent developments addressing these implications and trying to overcome the shortcomings of risk-assessment approaches, such as public engagement with GMOs and the ELSI (ethical, legal and social implications) on medical genetics and nanotechnology studies will be discussed. Finally, we will try to formulate some suggestions to help complementing existing risk-assessment studies with a more reflexive, socially-oriented approach. First, it is necessary to unpack the politics and ethics of a given technology, by addressing the emergence, the socio-technical networks, the power relationships and the economic interests that are tightly interrelated in the process of innovation and implementation. In this section – which tries to answer the question: what kind of future society is embedded in this technology? – the techno-social imaginaries and visions driving and underpinning technology innovation and implementation could be de-constructed and scrutinized, not only *per se* but also in relation to dominant socio-political imaginaries.

In the second section – which addresses the question: in what kind of society is this technology going to be implemented? – dominant practices, rules, formal and informal procedures and legislative gaps need to be explored. If risk-assessment procedures try to establish how safe is a technology, this approach rather tries to explore how "safe" is the context (Goven 2006b). In the third section, eco-social analysis should be integrated. GMOs affect agro-food production system, and have an impact not only on the environmental context in which they will be introduced but also in the way people feel and live and interact with this context. Social meanings, actions and relationships arise and are enacted around specific local environments and around the local understanding and framing of it. Changing these environments will inevitably change the socio-relational domains constructed around them. The potentiality of these changes, therefore, cannot be underestimated or neglected (Ferretti & Pavone 2009).

These three elements may help consolidating a more robust social assessment, which we define as an in-context trajectory evaluation. From this perspective, it emerges that there might be a number of socio-political reasons that support a moratorium on GMOs in Europe even if they come to be considered technically safe and ethically legitimate.

References

Felt U. et al. (2007) Taking European Knowledge Society Seriously. European Commission Working Document. EU Brussels.

Ferretti M.P., Pavone, V. (2009) What do civil society organisations expect from participation in science? Lessons from Germany and Spain on the issue of GMOs. Science and Public Policy 36 (4): 287–299.

Ferretti M.P. (2009) Risk and Distributive Justice: The Case of Regulating New Technologies. Science and Engineering Ethics, October 2009.
www.springerlink.com/content/a33006481 126ktk1, last accessed 12th February 2010.

Gasnier C. et al. (2009) Glyphosate-based herbicides are toxic and endocrine disruptors in human cell lines. Toxicology 262: 184–191.

Goven J. (2006a) Processes of Inclusion, Cultures of Calculation, Structures of Power: Scientific Citizenship and the Royal Commission on Genetic Modification. Science Technology Human Values 31: 565.

Goven J. (2006b) Dialogue, governance, and biotechnology: acknowledging the context of the conversation. The Integrated Assessment Journal – Bridging Sciences & Policy 6 (2): 99–116.

Heinemann J., Sparrow A., Traavik, T. (2004) Is confidence in the monitoring of GE foods justified? TRENDS in Biotechnology 22 (7): 331–336.

Jasanoff S. (2003) Technology of humility: citizen participation in governing science. Minerva 41: 223–244.

Jasanoff, S. (ed.) (2004) States of Knowledge: The Co-production of Science and Social Order. London and New York: Routledge.

Le Curieux-Belfond O. et al. (2008) Factors to consider before production and commercialization of aquatic genetically modified organisms: The case of transgenic salmon. Environmental Science & Policy 12: 170–189.

Levidow, L. (2007) European public participation as risk governance: enhancing democratic accountability for AgBiotech policy. Technology and Society 1: 19–51.

Levidow L. (2009) Democratizing Agri-Biotechnology? European Public Participation in Agbiotech Assessment. Comparative Sociology 8 (4): 541–564 (24).

Liberatore, A., Funtowicz, S. (2003) Democratizing expertise, expertizing democracy: What does it mean, and why bother? Science and Public Policy 30 (3): 146–150.

Mcnaghten P. et al. (2006) Nanotechnology, Governance, and Public Deliberation: What Role for the Social Sciences? Comparative Sociology 8 (4): 541–56.

Nowotny, H. (2003) Democratising expertise and socially robust knowledge. Science and Public Policy 30 (3): 151–156.

Seralini G. et al. (2009) How Subchronic and Chronic Health Effects can be Neglected for GMOs, Pesticides or Chemicals. International Journal of Biological Sciences 5 (5): 438–443.

Seralini G, Cellier, D., Spiroux de Vendomois, J. (2007) New Analysis of a Rat Feeding Study with Genetically Modified Maize Reveal Signs of Hepatorenal Toxicity. Archives of Environmental Contamination and Toxicology 52: 596–602.

Traavik T., Heinemann, J. (2007) Genetic Engineering and Omitted Health Research: Still No Answers to Ageing Questions. Penang: TWN.

Weingart P. (1999) Scientific expertise and political accountability: paradoxes of science in politics. In Science and Public Policy, 26 (3): 151–161.

Breckling, B. & Verhoeven, R. (2010) Implications of GM-Crop Cultivation at Large Spatial Scales. Theorie in der Ökologie 16. Frankfurt, Peter Lang.

The lack of regulation on GMO as one of the risk factors for biodiversity in a place of unique value – Example of the Lake Baikal Region

Natalia Sirina[a,b], Christiane Eschenbach[a], Wilhelm Windhorst[a] & Felix Müller[a]
(Ecology Centre, University of Kiel, Kiel, Germany; Irkutsk State University, Irkutsk, Russian Federation. – nata.sirina@gmail.com)

Introduction

The problem of wide spread of genetically modified organisms (GMOs) with impact on biodiversity caused by intentional or unintentional introduction of them is discussed controversially around the world (Richmond 2008; Ervin et al. 2003; Winter 2008). In developing countries and countries with transitional economics, such as Russia, these questions are especially alarming. The Baikal Natural Area (BNA) has a unique value because of high biodiversity and an enormous number of endemic species within the territory.

The goal of our research was to determine the risk of GMOs spread and potential impact on biodiversity of the Baikal Natural Area with regard to deficits in regulation.

Methods

A comprehensive literature study (including internet enquiry) and interviews with representatives of administration and specialists on legal regulation has been carried out. About 70 state level regulations were analyzed. The enquiry about regulation in the field of GMO in Russia to the Committee of the State Duma of Russia on Natural Resources, Nature Management and Ecology was made. Consultations about Russian regulation were held with the Director of the Institute of East European Law of Christian-Albrechts-University (Kiel) Prof. Dr. Alexander Trunk and the assistant professor of the Law Institute of Irkutsk State University Dr. Lisauskaite Valentina. In addition the literature about risks concerning GMO and regulation in the field of GMOs in other countries was surveyed.

Results and conclusion

Baikal Natural Area is a territory that was determined officially in the Federal Statute (FS) of the Russian Federation (RF) № 94 (FS 1999) as a territory including the Baikal Lake, the water fenced-off area adjoining the lake, catchment basin within the borders of the Russian Federation, adjacent to Lake Baikal specially protected natural areas and adherent area to the Baikal Lake in 200 kilometres wide in west and north-west from it (art. 2). This statute determines the special regime of protection of the Baikal Lake and of the surrounding area (BNA).

The Baikal Natural Area was separated into three ecological zones: the central ecological zone (with the strictest protection regime), the buffered zone and the ecological zone of atmospheric impact (Figure 1).

Fig. 1: Scheme of ecological zones of Baikal natural area with the Lake Baikal in the centre. Central zone: around the lake (dark grey); buffer zone: in the south-east (grey); zone of atmospheric influence: in the north-west (light grey). Source: http://www.geol.irk.ru/baikal/baikal.htm.

The area of BNA covers 386 000 km^2. There are more than 2000 species within the territory. The RF Ministry of Natural Resources and Ecology publishes information about the BNA as a fulfilment of FS № 94 e.g. Federal Target Programs, Territorial

Target Programs including all new regulations about BNA, maps, reports about Lake Baikal state and measures for its protection from 2003 annually as well as other information (Protection of Lake Baikal 2010). But up to now information about the biodiversity of the region is still not fully available.

There are some projects that focused on the biodiversity conservation of Lake Baikal. One of the most remarkable project was the Biodiversity Conservation Project in Russia with the Lake Baikal regional component (1997–2003), initiated by the Global Environmental Facility. The Baikal component of the project allowed developing strategic approaches to biodiversity conservation in the BNA. The Lake Baikal Protection Legislation Development Concept, elaborated under the Project, was supported by the State Duma, which drafted more than ten legal acts required for efficient implementation of the FS № 94 (Russian Federation 2003). One of them was adopted in 2001. It is the Governmental Regulation № 643 about specification assertion of a list of forbidden activities within the "Central Ecological Zone" of the BNA. In this act we can find restrictions of activities concerning GMOs. So, in the central ecological zone of BNA research and work connected with the use of genetically engineering technologies or conducting of any activity with biological objects, which leads to changes of their genetic structure, is forbidden.

In general, we found the regulation of GMOs and particularly with impact of environment within the Russian Federation to be in the slow development.

Russia ratified the Convention on Biological Diversity in 1995. According to this convention the Russian Federation regularly published reports about its activities on biodiversity conservation. By 2010, four reports about biodiversity conservation were published. But up to now Russia is among the few countries that have not ratified the Cartagena Protocol on Biosafety. So far, Russia has no commitment of biosafety risk assessment, management and monitoring of "living modified organisms ... that may have adverse effects on the conservation ... of biological diversity" (Cartagena Protocol, art. 1).

At the level of the Russian Federation there are no strict terms in the field of GMO and no clearly determined procedures for environmental risk assessment of GMO (FS 1996; GR 2001). For instance, in the main statute in the field of GMO (FS 1996) there are two close terms. One is "genetically modified organism" and the other is "transgenic organisms". Both, in genetically modified organisms and in transgenic organisms, genetic programs have been altered by means of genetic engineering techniques, but genetically modified organisms are supposed to be capable of multiplication or transmission of genetic material (art. 2). These terms may be confused in the course of application.

There is no term "environmental risk assessment" both in common Russian legislation and in legislation about genetically modified organisms. A definition of environmental risk exists, but it mentions just adverse effect for nature (FS 7, 2002, art. 1). And in the FS 86 describing "risk associated with the potentially harmful effects of genetic engineering activity on human beings and the environment" (art. 7) the risks were focused to health of people who work with these organisms within a closed system. So far, environmental risks of GMOs for the environment are not sufficiently covered.

Even though for the issue of introducing new species into the environment there are at least five laws: FS of the Russian Federation about government control in line of genetically engineering activity № 86 (1996), Governmental Regulation (GR) about the Federal Target Program "National system of chemical and biological safety of Russian Federation" № 791 (2008), GR about official registration of genetically-engineering modified organisms № 120 (2001), FS of the RF environmental protection law №7 (2002) and FS of the RF law of ecological expertise № 174 (1995) these acts probably are not enough to assure the biosafety of the country.

For protected areas there exists the FS of the RF about specially protected natural areas № 33 (1995). The statute defines rules for any activity in areas with special protection depending on status. The different regimes specified for the states natural reserves including biosphere reserves, state parks, natural parks, state natural sanctuaries etc. The strictest rules are applicable to the state's natural reserves, for example the prohibition of any activity that is in conflict with the protection of the natural territories in order to preserve biological diversity and support natural complexes and objects that are under protection in their natural state. But the case of GMO is not mentioned directly in this statute. So we should mention that BNA (Baikal Natural Area) includes five State nature reserves (Baikalo-Lenskiy, Baikalskiy, Bargusinskiy, Dzerginskiy, Sakhodninskiy) and three state parks (Pribaikalskiy, Zabaikalskiy, Tunkinskiy) with such strict regime.

At present, no GMO are registered for commercial harvesting in Russia. But there is a common register of cultivars that can be used only in the food processing industry and sale for population but are not allowed to be used for cultivation. 18 GM varieties (cultivars) e.g. potatoes (2), maize (9), and soy (4), are admitted for use in food production, feed, animal medicine and foodstuffs. A huge amount of foodstuff containing GMO are at the same register (Register 2010). There is also a separate register for feed for animals which include 209 kinds of feed contained GMOs (State feed register 2010).

From our results we may conclude that federal regulations for biodiversity protection in Baikal region exist since 1999. International projects as GEF help to strengthen this regulation including risks emerging from GMOs. But up to now the Cartagena protocol was not ratified, which regulates transboundary movement of GMO. There are some weaknesses of federal acts in the field of GMOs with regard to nature protection and biosafety strategy of the RF. This could make it possible that GMOs might spread unintendedly within the country in general and in the Baikal region in particular.

References

Ervin D.E., Welsh R., Batie S.S., Line C. (2003) Carpentier Towards an ecological systems approach in public research for environmental regulation of transgenic crops. Agriculture, Ecosystems and Environment 99: 1–14.
Cartagena protocol on biosafety to the convention on biological diversity. Montreal. 2000.
Governmental Regulation of the Russian Federation № 120 (2001) About official registration of genetically-engineering modified organisms.
Governmental Regulation № 643 (2001) About specification assertion a list of forbidden activities within the "Central Ecological Zone" of BNA.
Federal Statute of the Russian Federation № 7-FS (2002) About environmental protection.
Federal Statute of the Russian Federation № 86-FS (1996) About government control in line of genetically engineering activity.
Federal Statute of the Russian Federation № 94-FS (1999) About protection of Lake Baikal.
Protection of Lake Baikal (2010). http://www.geol.irk.ru/baikal/baikal.htm.
Register (2010) of production passed state registration (Given by Federal Service, including Authorities). http://fp.crc.ru/gosregfr/?oper=s&type=min&pdk=on&pril=on&text=%E3%E5%ED%ED%EE-%E8%ED%E6%E5%ED%E5%F0%ED%EE.
Richmond R.H. (2008) Environmental protection: applying the precautionary principle and proactive regulation to biotechnology. Trends in Biotechnology 26 (8): 460–467.
Russian Federation (2003) GEF Biodiversity Conservation Project: Outcomes and Prospects. Moscow, Publishing House of the Scientific and Training/Methodological Center. 48 pp.
State feed register (2010) obtained with derived from genetically modified organism. http://www.mcx-consult.ru/gosudarstvennyy_reestr_himicheskih_.
Winter G. (2008) Nature protection and the introduction into the environment of genetically modified organisms: risks analysis in EU multilevel governance. Review of European Community and International Environmental Law (RECIEL) 17 (2): 205–220.

Breckling, B. & Verhoeven, R. (2010) Implications of GM-Crop Cultivation at Large Spatial Scales. Theorie in der Ökologie 16. Frankfurt, Peter Lang.

GM maize and oil seed rape in Germany: Economic welfare losses from large scale adoption scenarios

Jan Barkmann[a], Manuel Thiel[a], Ludwig Theuvsen[a], Christiane Eschenbach[b], Wilhelm Windhorst[b] & Rainer Marggraf[a]
([a]Department for Agricultural Economics and Rural Development, Georg-August-Universität Göttingen, Germany; [b]Ecology Centre, Christian-Alberechts-Universität zu Kiel, Germany. – jbarkma@gwdg.de)

Introduction

We investigated the comprehensive economic effects of large scale cropping scenarios for Bt-Maize and HR-Oil Seed Rape (OSR) in Germany. While cropping of the Bt-Maize MON810 was allowed for until April 2009 in Germany, no HR-OSR variety is currently approved. Although approval procedures as well as their European Union legal framework are politically contested issues, studies that quantify the welfare economic effects of large scale cropping of GM maize and OSR for Europe's largest economy are missing.

Methods

Whether a specific environmental regulation is of comprehensive economic advantage or not, can be investigated by welfare economic cost-benefit analysis (CBA, Marggraf & Streeb 1997). In CBA, all advantages and disadvantages of a regulation are quantified in monetary units. Advantages in our case are higher producer profits or lower consumer prices. Non-market changes in consumer utility have to be included as well. One of these changes may stem from consumers with anti-GMO attitudes who are "forced" to consume food produced using GM crops because the availability of non-GM alternatives may be substantially reduced. Empirically quantifying this loss of consumer utility using a willingness-to-pay (WTP) survey is the main focus of this study.

WTP was assessed with a postal Choice Modelling (CM) survey (153 valid responses; for the CM method, see Louviere et al. (2000). WTP was calculated from statistically significant utility coefficients of an additive, linear conditional logit model estimated with NLOGIT 3.0.

In order to include a range of products from maize and OSR, respondents were confronted with cornflakes, canned sweet corn, vegetable cooking oil, margarine, milk, beef, and electricity generated from biomass. These products were offered at systematically differing prices and "content" of Bt-Maize or HR-OSR. GMP-"content" was reported as the result of laboratory analytics as well as the result of certificates of origin. Additionally, we included the possibility that GMP-free alternatives were available from regional, German, EU or global production.

We compared two large scale adoption scenarios to a base scenario without GM cropping. In the *70–50* scenario, Bt-Maize is grown on 70 % of the area (370.000 ha) in Germany threatened by the European Corn Borer (*Ostrinia nubilalis*). The macro-economic advantage is estimated at 10 €/ha/yr. In this scenario, HR-OSR is cropped on 50 % of current acreage (~ 750.000 ha; advantage: ~ 116 €/ha/yr). The *70–100*-scenario differs only for the proportion of HR-OSR grown.

If conventionally cropped areas are contaminated with > 0.9 % GMP, the crop must be sold at a lower price, e.g. as animal feed. For simplicity, we assumed that the price is lower by either 10 €/ha/yr or 116/€/ha (i.e. equivalent to the GMP cropping advantages in the two scenarios). The contaminated area was extrapolated from published detailed in Germany (Middelhoff et al. 2004; Schmidt et al. 2009).

Costs for an extended public GMP monitoring scheme were calculated as ~ 11m €/yr (Thiel & Marggraf 2009). Additional private separation and testing costs in the agricultural trade and production chain were estimated as 6 €/t/yr (Gavron & Theuvsen 2007).

Results

For food produced from and/or containing about 1 % GMP, respondents stated a statistically significant mean utility loss of 38 % of retail prices. This is a loss of 403 million to 574 million €/yr depending on the scenario. In the 70–100 scenario, the reduced regional/national availability of margarine and vegetable cooking oil from non-HR OSR results in an additional utility loss of 36 % of the retail price (324 million €/yr).

Discussion

As long-term environmental effects of large scale GM cropping are controversially debated, this source of costs or benefits was not included. Likewise, we did not include utility losses resulting from the fact that customers dislike beef and milk produced from GM crops because current consumer legislation in Germany does not require a respective declaration.

For both scenarios, CBA results in a clearly negative sum total. For each single € economically gained by lower production costs, 5 € direct costs and loss of consumer utility are incurred. Thus, legal approval of large scale cropping of Bt-Maize and HR-OSR is not indicated economically. CBA results reverse, however, if the loss of consumer utility was ignored.

References

Gawron J.-C., Theuvsen L. (2008) Kosten der Verarbeitung gentechnisch veränderter Organismen: Eine Analyse am Beispiel der Raps- und Maisverarbeitung. In: Glebe et al. (Eds) Agrar- und Ernährungswirtschaft im Umbruch. Münster, Münster-Hiltrup. 143–152.

Louviere J.J., Hensher D.A., Swait J.D., (2000) Stated Choice-Methods: Analysis and Applications. Cambridge, Cambridge University Press.

Marggraf R., Streb S. (1997) Ökonomische Bewertung der natürlichen Umwelt. Heidelberg, Spektrum.

Middelhoff U., Baumann R., Kessler M., Reiche E.-W., Rinker A., Tillmann J., Windhorst W. (2004): Teil D: Regionalstudie Schleswig-Holstein. In: Bbreckling B. et al. Generische Erfassung und Extrapolation der Raps-Ausbreitung. GenEERA. Abschlussbericht.

Schmidt G., Kleppin L., Schröder W., Breckling B., Reuter H., Eschenbach C., Windhorst W., Höltl K., Wurbs A., Barkmann J., Marggraf R., Thiel M. (2009) Systemic Risks of Genetically Modified Organisms in Crop Production: Interdisciplinary Perspective. GAIA 18(2): 119–126.

Thiel M., Marggraf R. (2009) Monitoring von gentechnisch veränderten Pflanzen – Konzeption und Kosten. In: Peyerl H. (Ed.) Jahrbuch der Österreichischen Gesellschaft für Agrarökonomie. Wien, Facultas. Band 18(1): 131–140.

Breckling, B. & Verhoeven, R. (2010) Implications of GM-Crop Cultivation at Large Spatial Scales.
Theorie in der Ökologie 16. Frankfurt, Peter Lang.

Risk Governance: Communication Strategies for Coexistence with GMOs (Genetically Modified Organisms)

Claudia Bethwell, Thomas Weith & Klaus Müller
(Leibniz Centre for Agricultural Landscape Research (ZALF), Dept. for Socio-Economics, Müncheberg, Germany – Claudia.Bethwell@zalf.de)

Introduction

The cultivation of GMO implies various impacts on whole landscapes. In accordance with the definition of OECD (2003) GMO cultivation is a systemic risk. The main reasons for that are the unpredictability of the extent of the potential harm and the probability of damage occurrence as well as the possibility of long-term and/or long-distance impact on land use systems (Renn & Keil 2008). As a result high ambivalence regarding the potentials and risks of GMO cultivation is to be noted. This ambivalence dominates the current discussion and perception processes.

Within Europe the Directive 2001/18/EC requires the coexistence of conventional farming, ecological farming and GMO-farming. Consequently strategic approaches are needed to handle coexistence, especially at the local and regional level. In theory a discourse strategy regarding the coexistence of different forms of production, including communication of risks and participative cooperation, constitutes an adequate way for handling the diversity in attitudes towards GMO and interests in specific regions (RRAC 2009). This reflects new forms of conflict handling and problem solving using a governance approach. A mix of informal and formal regulations and multiactor involvement are the central aspects (Assmuth, Hildén & Benighaus 2010; more general: Benz 2009).

This article presents the following points based on our experiences a) to which extent is it possible to develop and implement in reality such a cooperation and communication strategy b) what are workable approaches for developing rules of coexistence and c) how to define requirements for government agencies and policy makers at different levels. Our first experiences of such a discourse involving local administration, conventional and ecological farmers in the Brandenburg region Märkisch-Oderland will give indications for options of developing such a process as a general applicable tool.

The Brandenburg region Märkisch-Oderland: first experiences with GMO-farming

The model region Märkisch-Oderland is situated near Berlin in the eastern part of Brandenburg. This region is characterised by its circa 118,000 ha arable land and 9,600 ha maize cultivation (8 % of arable land) (Statistische Ämter des Bundes und der Länder 2009). Märkisch-Oderland is the centre of Bt-maize cultivation MON810 in Germany, with 550 ha Bt-maize cultivation compared to about 2680 ha Bt-maize cultivation in Germany as a whole in 2007 (MLUV 2008; BVL 2010).

Development and implementation of a communication strategy in the model region

In 2009 and 2010 the authors designed and developed a cooperation and communication strategy for the situation in Märkisch Oderland (Figure 1). This is an ongoing process. Based on the analysis of concerned local stakeholder groups a first workshop was realized to analyse the range of attitudes towards GMO. Subsequent expert interviews allowed a more detailed insight into the stakeholders' viewpoints (categories) and their expectations of problem handling with GMO. One important interim result indicated that the communication strategy should be focused on the following two objectives:
- defining "rules of coexistence" for the region, and
- elaborating "requirements to government agency and policy makers" of the region.

In consequence a second workshop was realized to develop these aspects. The following actors are involved in the ongoing communication process:
a. the head of the county administration,
b. the local environmental agency,
c. the local farming agency,
d. representatives of the regional farmers associations.

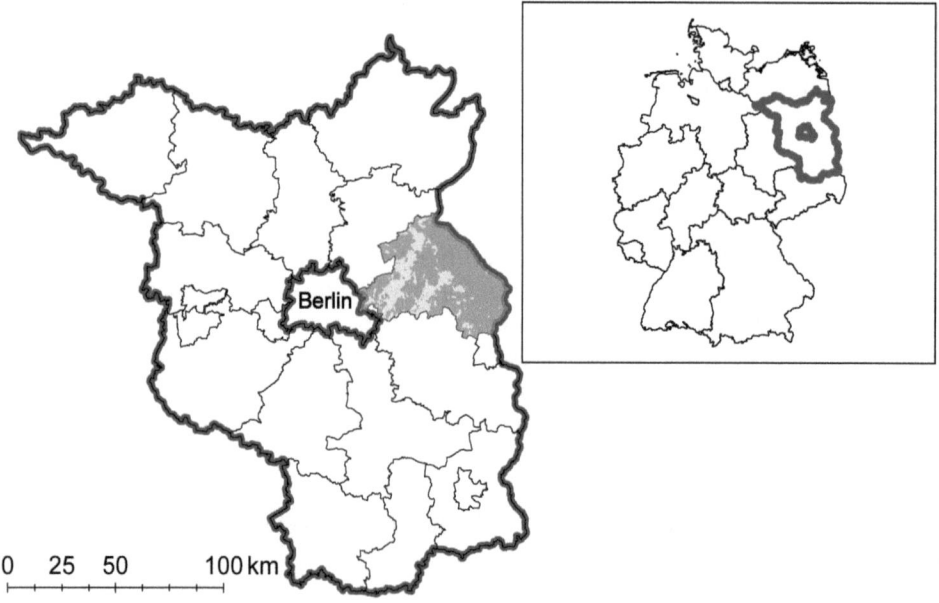

Fig. 1: Model region: The Brandenburg Region "Märkisch-Oderland". Grey: Märkisch-Oderland; dark grey: arable land

Focus: selected results of the expert interviews

Expert interviews with non-GMO-farmers and stakeholders of the local administration were used to explore the general attitudes towards GMO, the problems and conflicts of dealing with GMO, their position to the communicative approach, their suggestions and ideas for handling GMO problems. The interviews were analysed according to the qualitative content analysis from Mayring (2008). This method guarantees the inter-subjective and transparent analysis of the attitudes.

As a result the attitudes towards GMO varied between 'general refusal of GMO' or 'Bt-maize is not required in the region' (non-GMO-farmers) and 'responsible handling of GMO is needed' (local administration). If stakeholders accept the requirement of coexistence according to the European Directive 2001/18/EC and ask for the responsible handling of GMO and for the minimizing of local and regional conflicts, the development of a cooperation and communication strategy will be supportive. According to the local actors' opinion, the communication process should be focused on Bt-maize, because Bt maize is at present the only genetically modified plant that may be cultivated in the EU. In addition the capacity of every GMO is different and therefore the handling of every GMO (e.g. coexistence rules) should be specific.

The described examples of problems and conflicts in the region with regard to GMO-cultivation are the loss of confidence between GMO-farmers and non-GMO-farmers and limitation in help from neighbours up to legal disputes and political polarisations. The communicative approach is considered as a need for regional actors and a prerequisite for the coexistence between all forms of production. All experts announced: 'we need a communication strategy for the moment of repealing the ban on MON810'.

In the interviews the experts also suggested some "rules of coexistence within the region". This is what has been proposed: early communication with neighbours six months before sowing, enhanced distance between GMO and neighboured conventional-/eco-maize-fields, precautionary minimum distances on GMO farmland (also with support of a geographical information system), development of a best agricultural practice list dealing with GMO, exchange of arable land, preparation of a checklist for the GMO-farmers (e.g. to prove minimum distances; see above), improved farmers' qualification for dealing with GMO and improved knowledge about pest control for European corn borer (*Ostrinia nubilalis*), and introduction of or support measures for minimizing the pollen drift of GMO-maize.

Examples of suggested "requirements to government agency and policy makers demanded by the region" concern the general decision for or against GMO-cultivation, the predictability of legal decisions, the clarification of liability for seed producers and the question of the necessity of insurance coverage for GMO-farmers.

Perspectives

Currently the communication process is in progress in the region. Local actors stressed the necessity of further scientific assistance. A supportive tool for this in the future could be "Risk mapping" (Kropp, Beck & Engel 2007), an instrument for the visualisation of any communication processes. The tool allows for an internet based visualisation of risk related controversies as argumentation maps. It will be helpful for structuring the attitudes of the panellists and for the visualisation of the different attitudes. This will support one main aspect of the governance approach: a broader involvement of various regional and local actor groups (Assmuth, Hilden & Benighaus 2010).

For developing a complete set of rules for coexistence other national regulations, for example the Danish coexistence regulation, must also be analysed to improve the local approach in the Brandenburg region Märkisch-Oderland and to further discourse towards complex risk governance (De Marchi 2003).

References

Assmuth T., Hildén M., Benighaus C. (2010) Integrated risk assessment and risk governance as socio-political phenomena: A synthetic view of challenges. Science of the Total Environment, 408(18): 3943–3953.
Benz A. (Ed.) (2004) Governance. Regieren in komplexen Regelsystemen. Wiesbaden.
BVL (2010) Standortregister. http://apps2.bvl.bund.de/stareg_web/showflaechen.do.
De Marchi B. (2003) Risk governance Public participation and risk governance. Science and Public Policy 30: 171–176.
European Community (2001) Directive/2001/18/EC of the European Parliament and of the Council. Official J EU Commun 2001/18/EC: 1–64.
Kropp C., Beck G., Engel A. (2007) Risikokonflikte visualisiert. Ökologisches Wirtschaften 3/2007. München, Ökom Verlag.
Mayring (2008) Qualitative Inhaltsanalyse. Grundlagen und Techniken. Weinheim, Beltz.
MLUV Brandenburg (2008) Gentechnik in Brandenburg, Bericht 2008. Potsdam, MLUV.
OECD (2003) Emerging risks in the 21th century: An agenda for action (Final report to the OECD futures project). Paris. OECD.
Renn O., Keil F. (2008) Systemische Risiken: Versuch einer Charakterisierung. GAIA 17(4): 349–354.
RRAC (2009) A practical Guide to Public Risk Communication
    http://www.berr.gov.uk/files/file51458.pdf.
Statistische Ämter des Bundes und der Länder (2009) Statistik lokal. Daten für die Gemeinden, kreisfreien Städte und Kreise Deutschlands. Düsseldorf, IT.NRW.

EU and German law on coexistence: Individual and systemic solutions and their compatibility with property rights

Sarah Stoppe-Ramadan & Gerd Winter
(Research Centre for European Environmental Law, University of Bremen, Germany. – sramadan@uni-bremen.de)

Introduction

Coexistence of the cultivation of gm and non-gm crops is seen as a suitable way out of the clash of views on environmental or health risks of genetically modified organisms (GMO). It allows setting aside a clear risk based decision for or against GMOs, because all types of agriculture shall be given the possibility to exist side by side. While EU law is reticent[1] Germany promulgated quite elaborate rules on co-existence in its Genetic Engineering Act (GenTG) in 2008. However, the German co-existence measures leave certain problems unsolved.

Individual conflict resolution and systemic measures

The existing measures as laid down in the laws[2] and good practice rules[3] such as the register on GMO-release, prior information duties[4], distance rules for planting like crops, and liability for contamination of crops, all aim at solving conflicts between the individual landowners. They fail to recognize the systemic character of the problem. For instance, the rules on minimum distances have the effect that in regions with small agricultural plots much land must be reserved – and wasted – for puffer functions. The limited protection by § 36a GenTG (liability rules) also misconceives the systemic

---

1 Note: The article is based on a speech given at the GMLS II conference in Bremen, in March 2010. Therefore, it covers only the legal regulations being in effect until then. For the purpose of completeness, reference shall be made to the new Commission Recommendation on guidelines for the development of national co-existence measures to avoid the unintended presence of GMOs in conventional and organic crops of 13th July 2010, replacing Recommendation 2003/556/EC of 23rd July 2003 and the new proposal for an amendment to Directive 2001/18/EC. The first emphasizing inter alia the importance of making a clear distinction between health and environmental risks and promulgating the legitimacy of agricultural planning (including GMO-free regions). The latter proposes a new Article 26b to Directive 2001/18/EC leaving the decision whether or not GMOs may be cultivated to the Member States (restricted however only to such reasons being not already taken into account in the risk analysis).
2 For Germany see §§ 16a, 36a Act GenTG.
3 Regulation on generation of genetically modified plants (Gentechnik-Pflanzenerzeugungsverordnung – GenTPflEV) of April 7, 2008 (BGBl. I S. 655).
4 Cf. §§ 16b (5); 18 (2); 21 GenTG – duty to inform the authorities and participation of the public; as well as § 35 GenTG – duty to inform the injured party.

aspect of the problem, as the prerequisite of a neighbour (§ 906 II BGB) ignores more distant and non-agricultural actors. Beekeepers facing the problem of their honey getting contaminated with GMO pollen serve as an example.[5] Solutions to this problem could be landscape planning and the strengthening of voluntary agreements between the landowners.

a) Landscape planning
Taking Germany as a case the task of landscape planning is to describe the present and future natural areas and their uses, define objectives of nature protection and landscape preservation, evaluate the factual situation and elaborate measures in view of the defined objectives.[6] Measures of this kind may aim at the protection of nature against damage[7], but they may also aim at the preservation of the diversity, originality (Eigenart) and amenity of nature and landscape.[8]

The authors submit that the determination of areas where no GMOs shall be introduced can be one measure of this kind, i.e. one which aims at preserving diversity, originality and amenity. Such use of landscape planning would be compatible with the already cited Commission Communication where it is said that "measures of a regional dimension could be considered. Such measures should apply only to specific crops whose cultivation would be incompatible with coexistence, and their geographical scale should be as limited as possible."[9]

When considering landscape planning as a tool of coexistence it must however be noted that in most of the German Länder landscape plans are not legally binding. Nevertheless, even as a non-binding document it can provide guidance for the farmers.

b) Voluntary agreements
Voluntary agreements between the landowners are another method to arrange coexistence. They can not only be used to build up GMO free regions[10], but also to combine land into a GMO-region. However, these agreements are, as the name indicates, only voluntary. It is hardly probable that a landowner who promises not to release GM plants

---

5 In the case of Brandenburg for example, pollen of the maize MON 810 was found in the honey. The bee-keeper applied for appropriate measures of the competent administrative body to avoid contact of his bees with the said maize pollen. The judge agreed in the fact that bee-keepers come into a hopeless situation if the deliberate release of GMOs is expanding; however it is for the legislator to find a solution to this problem.
6 § 9 (3) Federal Nature Protection Act (Bundesnaturschutzgesetz – BNatSchG).
7 § 9 (3) No. 4 (a) BNatSchG.
8 § 9 (3) No. 4 (f) BNatSchG.
9 Commission Recommendation of July 23, 2003 on guidelines for the development of national strategies and best practices to ensure the coexistence of genetically modified crops with conventional and organic farming (notified under document number C(2003)2624, Introduction 2.1.5.
10 See, for instance, the Milchwerke Berchtesgadener Land Chiemgau eG, http://www.molkerei-bgl.de/files/assets/downloads/GVO_Frei.pdf, last visited on March 9, 2010; or AbL Baden-Württemberg, Aktionsbündnis Gentechnik-freie Region Oberrhein, Bioland Baden-Württemberg, Demeter Baden-Württemberg, Evangelisches Bauernwerk in Württemberg, Naturland Süd-West, Nürtinger Bündnis für gentechnikfreie Landwirtschaft und Lebensmittel, Verband Katholisches Landvolk e.V., http://www.demeter-bw.de/gentechnik_buendnis.php, last visited on 9 March 2010.

will agree to sanctions or to accept a servitude which would make his commitment binding also on subsequent owners of his land.

c) Binding administrative regulation
As neither landscape planning nor voluntary agreements lead to binding rules of coexistence we suggest that the matter should be regulated by administrative law, for instance by the introduction of binding agricultural planning in relation to the use of GM plants.

Problems of encroachments on constitutional rights

Any measure solving the conflict between GM- and non-GM farmers, may it address the individual or the systemic conflict, raises questions of compatibility with the constitutional guarantee of property.

Concentrating ourselves on the property rights of the GMO-farmer (under Art. 14 GG), his property guarantees could be affected because he must register his crop, keep distances to other crops, compensate damage, obey agricultural planning, etc.

The protection scope of the property guarantee encompasses the choice of seeds the farmer wishes to sow. Therefore, to hinder him from doing so freely (by the above mentioned regulations on coexistence) is an intervention into his right. However, property is subject to limitations defined by law. Therefore, an intervention can be justified by law. Even though the legislator has broad discretion to fix such limitations and even define the extension of protected property, they may not destroy the essential core of the property principle, must pursue a public interest, and must observe the principle of proportionality.

Subsuming the case under these rules it is clear from the outset that the essential core of property is not touched if a farmer cannot release GM plants. As to the public interest it can be found in the objective of ensuring the coexistence of different ways of agriculture. There is a public interest to allow non-GM farmers to carry on and provide non-GM food for those consumers who have a preference for non-GM food. This public interest is even backed by constitutional rights of property of non-GM farmers to be protected against contamination, and maybe also by personal freedoms of consumers to have a free choice of goods.[11]

Proportionality further demands that no less restrictive measure could have been used to achieve the objective. The measures addressing the individual conflict between farmers appear to be less restrictive than binding agricultural planning. However, it is open to

---

11 It is controversial if the consumers' free choice is constitutionally guaranteed. The right to be considered would be, under the German constitution, the right of free development of personality (Article 2 (1) GG). Certainly, this right would not embrace the right of choice between 100 types of perfume. However, there is a right to be served with essential goods ranging from food to services and cultural assets. Maybe, the general availability of non-GM products could be understood to belong to this realm of essential goods.

doubt if these measures effectively ensure coexistence. For instance, given the fact that pollen is transported over long distances, land corridors around GMO-fields may fail to separate productions in the long run. Therefore, also binding measures would pass the test of proportionality.

Breckling, B. & Verhoeven, R. (2010) Implications of GM-Crop Cultivation at Large Spatial Scales.
Theorie in der Ökologie 16. Frankfurt, Peter Lang.

Legal implications of the step-by-step principle on risk assessment of GMOs

Caroline von Kries[a] & Gerd Winter[b]
([a]Attorney at Law, Freiburg, Germany; [b]Research Centre for European Environmental Law, University of Bremen, Germany. – caroline.v.kries@gmx.de)

Introduction

The step-by-step principle was introduced into EU legislation on genetically modified organisms (GMOs) as a means to cope with uncertainty about environmental risks from the release of GMOs into the environment. The approval process is orientated along the stepwise reduction of containment thus adopting a precautionary approach towards the risks of GMOs. The stepwise scaling up of release shall keep pace with the gradual generation of risk related knowledge. This paper strives to clarify the meaning, legal status and practical importance of the principle.

'Step-by-step' in the relevant legal texts

The core formulation of the principle is contained in considerations (24) and (25) of Directive 2001/18/EC:

> "(24) The introduction of GMOs into the environment should be carried out according to the 'step by step' principle. This means that the containment of GMOs is reduced and the scale of release increased gradually, step by step, but only if evaluation of the earlier steps in terms of protection of human health and the environment indicates that the next step can be taken.
>
> (25) No GMOs, as or in products, intended for deliberate release are to be considered for placing on the market without first having been subjected to satisfactory field testing at the research and development stage in ecosystems which could be affected by their use."

Directives must be transposed into Member State law allowing them some legislatory discretion. In German law, for example, 'step-by-step' is framed in rather broad language requiring that the risk assessment "shall be based on experiences made on previous steps".[1] Contrastingly, the Austrian Gentechnikgesetz (Genetic Engineering Act – GTG)

---

1 Genetic Engineering Procedure Regulation (Gentechnikverfahrensverordnung – GenTVfV.

establishes 'step-by-step' as a binding precondition of authorisations. See Sec. 3 (3) which reads:

> "The release of GMOs may only be performed step by step meaning that the containment of the GMOs may stepwise be unclenched and the scale of release only be increased if the assessment of the earlier step indicates that the next step is compatible with the precautionary principle."

A similar rule is contained in Art. 6 (2) of the Swiss Gentechnikgesetz (Genetic Engineering Act) (Errass 2006: 170).

Scope of application and legal status of the step-by-step principle

The principle is applicable to all authorisation procedures concerning the experimental release and the placing on the market of GMOs which result in the subsequent introduction of GMOs into the environment. By reference to Regulation (EC) 1829/2003 of 22 September 2003 it is also applicable to the authorisation procedure concerning the placing on the market of genetically modified food and feed.

Considering that the principle is part of the preamble but not of the working text of Directive 2001/18/EC it is (in its quality as EU law) not a self-standing prerequisite of authorisations (Administrative Court Berlin 1995). It nevertheless has a legal value. First, it is a general principle explaining the overall philosophy for the introduction of GMOs into the environment, i.e. the stepwise reduction of containment with accumulation of knowledge. Second, it is an interpretation guidance for the subsequent provisions thus helping to specify the scope of documents to be submitted and the understanding of the material yardsticks of risk assessment (Palme/Schlee 2009: 102). Member states are however free to transform this guidance into a binding prerequisite – as Austria did.

Basic information requirements for approval

According to Directive 2001/18/EC and Regulation (EC) 1829/03 the applicant for approval bears the burden of proving that no adverse effect will be caused by the release. This means that the applicant must if needed conduct certain tests, be it by his own initiative or upon request by the authority. There are four requirements which the applicant must fulfil in that respect, and which if unfulfilled allow the authority to reject the application. These can also be regarded as minimum postulates of the step-by-step principle:

- Submission of data on the parent organism, the recipient organism, the GMO and the effects of the GMO on human health, plant and animal health and the environment as listed in Art. 5 (3) and Annex III B Directive 2001/18/EC

- Submission of information identifying and characterising hazards and the likelihood of their occurrence as part of the environmental risk analysis (e.r.a.) as expounded by Annex II Dir 2001/18/EC
- Execution and submission of additional tests on demand of the authority if there is grounded hypothesis for an adverse affect on health or the environment.
- Submission of uncertainty analyses on all test results concerning health and environmental risks.

Meanings of 'step-by-step'

There are different occasions during the application phase when reference to the step-by-step principle may be of importance.

- The step-by-step principle shall ensure that the **state of science and technology** is developed throughout the relevant steps such that an adequate risk assessment can be made. If further previous scientific investigation is necessary for the risk assessment at the stage of the administrative decision and if the investigation can be conducted at a previous stage, the authorisation of deliberate release or placing on the market must be denied (Errass 2006: 170).
- Risk assessors and competent authorities frequently argue that additional studies are not needed because the assumed **risk is 'negligible' or 'tolerable'**. The authors suggest that this argumentation should only be accepted if it is balanced by a benefit. Such benefit is scientific progress at the stage of small or large scale release and more environmentally friendly agriculture at the stage of placing on the market (Winter 2008).
- The operator can ask to be freed from certain tests if the **relevant knowledge is available from other studies**, under the condition that the other studies are valid and reliable, on the same GMO and under like conditions (Palme 2009: 102).
- The authority is entitled to ask the operator to **submit all risk information obtained from earlier steps**. This includes both positive and negative test results. As the risk assessment shall explore also unpredicted effects it is not bound to risk hypotheses in respect of existing knowledge.
- The competent authorities are entitled to **generate and use their own administrative knowledge** in order to elaborate their own views. In this case the notifier must be given opportunity to comment on the information before the decision is taken.
- The authority is entitled to use risk **information it has obtained from other applications** if the information is not confidential (i.e. does not cause economic harm if disclosed) and if the other applicant gives his consent. If the consent is refused the applicant must produce the data anew (Art. 6 (4) and 25 (2) Directive 2001/18 EC).
- Within certain limits the **monitoring can be seen as an additional step** within the step-by-step approach. The objective of monitoring, be it case-specific monitoring or general surveillance, is to identify effects of the GMO(s) on human health or the environment which have not been discovered at the stage of the environmental risk

assessment. For deliberate releases this is generally expressed in Art. 6 (2) (v) Directive 2001/18/EC. The possibility of requiring monitoring does however not allow the competent authority to shift the testing of grounded risk hypotheses to the monitoring stage. As already mentioned it must deny authorisation if the test could have been performed within previous steps.
- Authorities do not have powers to command **tests that generate information needed at the subsequent step**. However, this does not hinder authorities to require, on the subsequent step, risk information which should have been generated on previous steps.

Conclusion

Although 'step-by-step' is not a precise legal rule it does have legal meaning as a principle guiding the risk assessment and management of GMO introduction into the environment. Assuming a process of gradual reduction of containment and scaling up of release ranging from closed systems via experimental release to cultivation seed the step-by-step principle requires that the knowledge on environmental risks of GMOs should be generated on stages previous to the ones where the risk can result in damage.

References

Legislation
Directive 2001/18/EC of the European Parliament and of the Council of 12 March 2001 on the deliberate release into the environment of genetically modified organisms and repealing Council Directive 90/220/EEC, OJ L 106, 17.4.2001, p. 1–39.
Regulation (EC) No 1829/2003 of the European Parliament and of the Council of 22 September 2003 on genetically modified food and feed, OJ L 268, 18.10.2003, p. 1–23.

Court decisions
Administrative Court Berlin, Judgement of 18 July 1995, in: Eberbach/Lange/Ronellenfitsch (eds.) Gentechnikrecht/Biomedizinrecht, Entscheidungssammlung, Nr. 7 zu § 16 GenTG.

Literature
Errass C. (2006) Öffentliches Recht der Gentechnologie im Außerhumanbereich, Bern, Stämpfli.
Palme C., Schlee M. (2009) Gentechnikrecht, Wiesbaden (Kommunal- und Schulverlag).
Winter G. (2008) Nature Protection and the Introduction into the Environment of Genetically Modified Organisms: Risk Analysis in EC Multilevel Governance. RECIEL 17 (2): 205–220.

Breckling, B. & Verhoeven, R. (2010) Implications of GM-Crop Cultivation at Large Spatial Scales. Theorie in der Ökologie 16. Frankfurt, Peter Lang.

Dead end developments – lessons learned from unsuccessful GMO

Broder Breckling
(Chair for Landscape Ecology, University of Vechta, Germany –
bbreckling@iuw.uni-vechta.de)

Introduction

In 1994, the "FlavrSavr" tomato was commercialised as the first genetically modified food for human consumption on the US market – and it turned out to be a failure. Was it a rare exception contrasting an overall successful trend in the development of genetically modified organisms – or does it represent a typical situation? To obtain a realistic impression about the technology, it is necessary not only to count the cases where a GMO so far met the expectations of economic gains. It is also required to see the cases that did not work. Since the reasons that rendered GMO unsuccessful are only partly documented in the scientific literature, a targeted search was required. A GMO is considered as a dead end development if it remained far below the initially expected application potential and either never became commercialised or was withdrawn from the market after economic damage occurred.

The tomato disappointment was documented by Martineau 2001, who participated in the development. Durable and aromatic fruits were the promise – resulting from blocking the expression of an enzyme involved in natural fruit maturation: The genetic information for polygalacturonase was inserted in reverse order, leading to a suppression of the formation of this enzyme. Only three years later, the transgenic trait was abandoned. Conventionally bred varieties proved to be superior. Since then, a large and partly unknown number of other GMO failed in practice. For an estimation how realistic future promises could be, it is quite useful to look back what happened to some of the previous attempts and promises.

The question is, whether there are underlying pattern which allow some indication of trends to expect by understanding the causes for cancelled developments. This could help avoiding unnecessary investment of both, public and private funds. We put forward the hypothesis, that there is not a typical threshold level that separates successful and disappointing traits. The reasons for failure cover the complete spectrum of possibilities, involving technical, genetic, ecological and agronomic as well as social and economic factors. Examples of failure on the different levels are given.

Prominent cases of unsuccessful GMO

After more than fifteen years of commercial use, the number of profitable GM traits is still limited compared to the numerous attempts in research and development. It mainly comprises herbicide resistance and insect toxic Bt plants. A small selection of the structurally more diverse abandoned cases is presented in reverse temporal order.

High Lysin maize LY038 (market introduction cancelled in 2009)
The purpose of the genetic modification was to increase the nutrition value of maize as animal feed by raising its content of the amino acid lysine. This was achieved by inserting a gene from *Corynebacterium glutamicum*. LY038 maize, developed by Renessen, a Monsanto / Cargill subsidiary, had gained admission for commercialisation in the USA. On other globally relevant markets, e.g. Canada, Japan, Australia and New Zealand the variety was also admitted for feed. The GM Crop database of the Centre for Environmental Risk Assessment (2008) lists notification documents. In a statement of 2005, Heinemann had raised scientific safety concerns. The developers applied for admission also in the European Union. The European Food Safety Authority (EFSA) requested additional data from the company to demonstrate the safety of the product. The applicant was asked to submit data on feeding experiments using processed (cooked) material. So far, only raw material had been tested. Cooked maize has a higher relevance to estimate safety for human consumption, e.g. in case of unintended impurities in processed food. The Bioscience Resource Project (2009) explained: cooking of food with high amounts of free lysine can lead to reactions of the amino acid with sugars, forming 'advanced glycoxidation end-products' (AGEs). They are known to be linked to the etiology of several human diseases, thus a relevant safety issue.

Instead of performing the animal feeding tests and providing the data, the applicant withdrew the application and insisted on a complete return of all documents from the authority. Thus, the safety concerns could not be investigated further (Bottemiller 2009). The withdrawal was said to be due to economic reasons. So, the request of data from a feeding experiment must have made a trait already admitted on major markets economically non-viable? Alternatively, it can be speculated whether a public controversy on health risks after admission might have been an issue.

Weevil resistant pea (development was stopped in 2005)
The purpose of the genetic modification was to protect pea cultivation in Australia against a weevil (*Bruchus pisorum*) which causes considerable damage. A bean-specific anti-nutrient was inserted to pea. The work goes back to approaches made during the 1980ies (Schroeder et al. 1995). The modification was tested to be efficient. Also feeding studies did not indicate problems with product safety. However, additional tests with the purified protein before commercialization caused the unexpected result that mice showed symptoms of inflammation of the lungs when exposed to the substance resulting from genetic modification unlike when exposed to the conventional substance (Young 2005). Further investigations revealed that even though the amino acid sequence of the protein was identical in pea and in beans, the end products differed.

Many proteins receive sugar molecules as attachments. The enzymes which catalyse the connections are to some extent species specific. Such an altered glycosylation pattern occurred and made the transgenic substance more immunogenic. After ten years of development in Australia, the trait was stopped due to these unexpected physiological effects.

Herbicide resistant wheat (market introduction failed in 2004)
Monsanto offers a number of plants which carry a transgenically mediated tolerance to the herbicide Round-up. In the USA, Argentina and Brazil, herbicide resistant soy beans have a very high market share and are commercially successful. To reduce the effort of weed control, also wheat varieties with this resistance were developed to the level of market introduction. In the public, there were environmental concerns e.g. that relying massively on a single herbicide may increase the spread of herbicide resistant weeds (Gurian-Sherman 2003). Crucial for the development, however, became objections from farmers and their organizations in Canada. The Canadian Wheat Board declared that for the moment no way was seen for an efficient segregation of GM and conventional wheat which might cause losses on export markets. It was pointed to consumer resistance in many countries. Monsanto declared to postpone the market introduction of the variety (Nickel 2009).

Triffid Flax (commercialization stopped in 2001 but "re-emerged" as an impurity in 2009)
At the University of Saskatchewan a genetically modified flax was developed with a resistance against ALS. The idea was to allow for higher levels of agrochemical residues from previous crops in the soil – to an extent that conventional flax does not tolerate. Admission was gained in the USA and in Canada. Growers, however, massively resisted – largely for crop purity reasons. Shortly after notification, the variety admission was withdrawn. It had been commercially available for a very short time only (Warick 2001). Almost ten years later, in 2009, GM impurities of Triffid flax in exports from Canada were discovered in Germany. For GMO without admission zero tolerance applies in most countries. Subsequent investigations in various other states around the globe found the same contamination in Canadian flax exports (GM Contamination Register 2010, Schmidt & Breckling, this volume). Many export charges had to be withdrawn leading to high economic losses. How Triffid became a widespread contamination so many years after withdrawal from the market remains an open question.

Interesting is also the name of the variety. It is identical with a 1960ies fiction movie about an imaginary carnivorous plant (Figure 1). The developer, who chose the name, Alan McHughen, served as the founding president of the International Society for Biosafety Research (ISBR). He is currently its treasurer (University of California, Riverside, 2009, spin profiles 2010). No information was found whether he or who else was held liable for the losses that occurred during the attempt to eliminate the widespread impurities from current cultivation.

Fig. 1: Adversisment poster of the triffid movie (source: http://en.wikipedia.org/wiki/File:Dayofthetriffids.jpg)

Starlink maize (withdrawn 2000)
This case is probably one of the most dramatic in the history of genetic modification (Bratspies 2003). It caused compensation payments which are estimated to sum up to several hundred millions of dollars (FOEE 2000). The Starlink maize variety, developed by Aventis Crop Science, contained a gene from *Bacillus thuringiensis* to make it toxic for the corn borer and other lepidopteran pests. The toxin was more stable than in other commercially available Bt maize variants. The degradation of the protein in digestive fluids was relatively slower. Admission was gained in the USA for feed only. Though farmers were informed and trained to segregate the maize from harvests for human consumption, Starlink traces were found in many food items with maize ingredients (Cox 2000). Cross-pollination between crops as well as mixing of harvests was likely to play a role in having brought up this impurity. Taco Bell taco shells were the first food in which the unapproved variety was discovered. Later, it was also found in maize exports and food aid in many countries. Aventis had to pay the highest amount of compensation so far. In 2002, Aventis sold its Crop Science branch to Bayer.

The soil bacterium Klebsiella planticola (development stopped in 1999)
Klebsiella planticola can naturally grow in the rhizosphere of plants. A modified strain was engineered to speed up the fermentation of plant residues (Ingham 1999). In biosafety experiments, it was found, that unlike the wild type, the modified strain was able to damage crop plants like wheat when associating to crop plant roots (Holmes & Ingham 1999). After discovering this unexpected effect, development was stopped. Since then, the genetic modification of bacteria for commercial release to the open environment received little interest, if any.

Nutrient enriched soybean (1996 cancelled in a late stage of development)
One of the early ideas of genetic modification was an attempt to improve the nutrition value of soy beans for human consumption. It was intended to increase the content of the amino acid methionine, which is relatively rare in conventional soy. A protein with high methionine content from Brazil nut was transferred. Unknowingly, a major aller-

gen was used. Persons with Brazil nut allergies showed strong symptoms when being exposed to the GM soy bean (Nordlee et al. 1996). The case sparked public debates whether it is useful to introduce compounds to other plants to improve the quality. Since then, an estimation of potential new allergens contained in GM plants became a standard.

Other cases

So far, cases were listed that received larger attention at the respective time. There are other cases which were not that widely discussed showing the relevance of additional factors that made a GMO unsuccessful.

Virus resistant sweet potato (Kenya)
For several years Florence Wambugu, a Kenyan molecular biologist and co-operant to Monsanto, toured the world to advertise GM to feed the poor in Africa (spinprofiles.org). About 6.000.000 $ of research funds were used for development of a virus-resistant variety of sweet potato. At the end of the funding period the development was abandoned. The transformation did not make the variety more resistant than conventionally bred crops. In Uganda, a neighbouring country, a conventional variety was bred in less time with better resistance (deGrassi 2003; New Scientist 2004).

Bt sunflower.
Snow et al. (2003) published a paper reporting empirical evidence that wild and weedy sunflower populations might benefit from hybridising with transgenic Bt sunflowers. Under pest pressure, the weedy transgenic hybrids produced up to 50 % more seeds than the wild type. This caused the preoccupation that the transgene might escape cultivation and lead to weed problems in other crops or become invasive. Bt sunflowers were not introduced so far.

Pea for pig vaccination
To compensate for antibiotics which are phased out in industrial livestock farming, it was intended to develop a pea which expressed an antibody against a relevant infectious bacterium in piglets. The development was given up because of doubts in efficiency, because of limited interest of investors and because of public criticism. In 2008, the producer Novoplant (Gatersleben, Germany) went off the market after field releases and feeding tests ended (Biotop 2008; Bauer 2007).

Zeaxanthin potato
Using public funds under the framework of biosafety research, a potato with an increased content of Zeaxanthin was developed and tested at the University of Weihenstephan (Munich). The attempt to facilitate nutrient enhanced potato chips ended after the funding period was over in 2008. No investors were interested to commercialise the product. (http://www.biosicherheit.de/de/kartoffel/ inhaltsstoffe/467.doku.html).

Conclusions: Reasons for canceling development or market withdrawal – implications for some still ongoing developments

The presented cases show, that the reasons for the failure of transgenic organisms are diverse. The cases cover practically all stages along the development pathway starting with unreasonable ideas which did not get far beyond the first laboratory studies and ending after market introduction. Major reasons involved were in particular:

Difficulties during laboratory- or field-testing:
- the construct did not work as intended
- potential damage to other crops was anticipated
- the product was allergenic or had unexpected compositional properties
- unexpected non-target or environmental effects occurred
- low market potential was anticipated;

Issues relevant during or after market introduction:
- strong resistance of potential growers or lack of acceptance e.g. because of crop purity concerns (admixture avoidance)
- poor quality of the product
- no consumer acceptance, unforeseen lack of economic viability
- unexpected dispersal after commercialisation beyond of what the admission allowed
- liability problems.

Outlook on problematic cases still in development.

Considering past experiences, some currently still relevant cases can be identified where the prospects could be limited.

Starch modified AMFLORA potato (European Union)
In 2010, Amflora, a BASF development, was admitted for commercial cultivation in the EU. The variety is to be used for industrial purposes (Williams 2010). Only traces are allowed in food. The potato produces one type of starch. A second starch type normally contained in potato is blocked. This reduces purification efforts in industrial processing. Meanwhile conventional varieties with similar characteristics in starch composition exist. The question arises, why the society should take the efforts required for trait segregation after the intended quality was achieved also by conventional breeding. Therefore it may be assumed that the future of this modification could be limited.

Herbicide resistant creeping bentgrass (USA)
Herbicide tolerant golf courses are considered desirable by a developer in the USA. The tolerance is against the same herbicide as used for crop plants. Creeping bentgrass also occurs as a weed and is a target of herbicide application in crop management. The grass is wind pollinated and seeds are light and have a high dispersal potential (Reichman et al. 2006). The risk is very high that HR bentgrass would invade crop land. This appears as an efficient way to make other commercial GM herbicide resistance traits in crop

plants obsolete. Nevertheless, considerable amount of investment of real money went into this idea.

Brinjal (India)
The egg plant and its wild relatives are native to India and play an important role in the traditional diet. It was considered to be a good idea to create a Bt variety though a comparable situation exists as with Bt sunflowers. The genetic modification would be difficult to control and to limit its dispersal because of potential gene escape and traditional small-scale and subsistence agriculture (Assam Small Farmers' Agri Business Consortium 2006). As a response to the heated public controversy in India, the government halted admission.

The considered cases support the conclusion, that commercially successful GMO require very high levels of safety testing together with an explicit acceptance of the society. Less controversial applications are those under containment conditions as for many micro-organisms used to synthesise pharmaceutical ingredients. Since an "all inclusive" view in development anticipation seems not to be a common place at the moment, we can expect a continuing number of failures. To avoid them, not only technical feasibility but the entire range of issues along the development pathway would have to be considered in advance – from molecular issues, physiological and ecological implications to socio-economic and cultural issues. When used in agriculture, GMO interact with very complex systems. While a large number of feasible ideas end already at the lab stage, it was shown that even after market introduction commercial disasters can occur. If expanding the time-scale of consideration also the currently economically viable GMO (in particular the currently marketed herbicide resistant crop plants and Bt plants) might to lose economic viability on the long run because of the emergence of resistant pests.

Is it reasonable to expect that anticipation will improve so far that dead end development of GM crop plants becomes exceptional? It seems that the frequency of unexpected problems is not decreasing. Genetic modification seems to share this with developments in other fields of technology where former achievements become obsolete. The difference with GMO is that, unlike in other technologic products, after deliberate release GMO can reproduce, multiply, disperse and evolve.

References

All quoted internet sources were accessible in May 2010.

Assam Small Farmers' Agri Business Consortium (2006) Briefing paper on Br Brinjal. http://assamagribusiness.nic.in/bt_brinjal_briefing_paper.pdf.
Bauer A. (2007) Genbank Gatersleben: Gentechnik oder genetische Ressourcen? Umweltinstitut München. http://www.umweltinstitut.org/download/gatersleben_hintergrund.pdf.
Bioscience Resource Project (2009) Transgenic High-lysine corn LY38 withdrawn after EU raises safety questions. http://www.bioscienceresource.org/news/article.php?id=43.

Biotop (2008) Novoplant-insolvenz: Aus für die Pharma-Erbsen.
  http://biotop.de/news/article+M51ef5393c34.html.
Bottemiller H. (Nov 2009) GM Corn pulled due to food safety concerns.
  http://greenbio.checkbiotech.org/news/gm_corn_pulled_due_food_safety_concerns.
Bratspies R. (2003) Myths of voluntary Compliance: Lessons from the StarLink Corn Fiasco. Michigan State University DCL College of Law, Public Law and Legal Theory Working Paper Series Research Paper No. 01-07. http://papers.ssrn.com/sol3/Delivery.cfm/SSRN_ ID421700_code030808500.pdf?abstractid=421700&mirid=3.
Center for Environmental Risk assessment (2008) GM Crop database REN-ØØØ38-3 (LY038). Food Standards Australia New Zealand (2006) Final Assessment Report Application A549 Food derived from high lysine corn LY038.
  http://cera-gmc.org/docs/decdocs/07-219-001.pdf.
  Health Canada (2007) Novel Food Information High Lysine Corn Ly038.
  http://cera-gmc.org/docs/decdocs/07-075-001.pdf.
Cox J. (2000) StarLink fiasco wreaks havoc in the heartland Developer wants EPA to approve seed for food supply. USA Today 27.10.2000.
  http://www.netlink.de/gen/Zeitung/2000/001027.html.
deGrassi A. (2003) Genetically modified crops and sustainable poverty alleviation in Sub-saharan Africa. An assessment of current evidence. Third World Network – Africa.
  http://allafrica.com/sustainable/resources/view/00010161.pdf.
FOEE (2000) The Starlink Scandal FOEE Biotech Mailout 6(7).
  http://www.foeeurope.org/GMOs/publications/vol6no7.pdf.
GM Contamination Register (2010) FP967 ('Triffid') flax has been gown illegally in Canada and exported around the globe.
  http://www.gmcontaminationregister.org/index.php?content=nw_detail1.
Gurian Sherman D. Roundup Ready Wheat – an overview based on advancements in the risk assessment of genetically engineerd crops.
  http://www.cspinet.org/biotech/reports.html.
  http://www.cspinet.org/biotech/RRwheat_paper.pdf.
Heinemann J. (2005) Submission on Application A549 Food derived from High Lysine Corn Ly038: to permit the use in food of high lysine corn. Submitted to Food Standards Australia/ New Zealand (FSANZ) New Zealand Institute of Gene Ecology, University of Canterbury, 69 pp. http://www.testbiotech.org/sites/default/files/LY038-highlysinecorn-INBIsubmission _Heinemann.pdf.
Holmes M., Ingham E.R. (1999) Ecological effects of genetically engineered *Klebsiella planticola* released into agricultural soil with varying clay content. Appl. Soil Ecol. 3: 394–399.
Ingham E. (1999) Good Intentions and Engineering Organisms that Kill Wheat. By Elaine Ingham, Oregon State University. http://www.greens.org/s-r/18/18-14.html.
Martineau B. (2001) Food fight: The short, unhappy life of the Flavr Savr tomato. The Sciences 41 (2): 24–29.
New Scientist (2004) Monsanto failure. A showcase project to develop a genetically modified crop for Africa has failed.
  http://www.newscientist.com/article/mg18124330.700-monsanto-failure.html.
Nickel R. (2009) Canadian wheat board cautious about GM wheat.
  http://www.reuters.com/article/idUSTRE54E59X20090515.
Nordlee J.A., Taylor S.L., Townsend J.A., Thomas L.A., Bush R.K. (1996) Identification of a Brazil-Nut allergen in transgenic soybeans. The New England Journal of Medicine 334: 688–692. http://content.nejm.org/cgi/content/full/334/11/688.

Reichman J.R., Watrud L.S., Lee E.H., Burdick C.A., Bollman M.A., Storm M.J., King G.A., Mallory-Smith C. (2006) Establishment of transgenic herbicide-resistant creeping bentgrass (*Agrostis stolonifera* L.) in nonagronomic habitats. Molecular Ecology 15: 4243–4255.

Schmidt G., Breckling B. (this volume).

Schroeder H.E., Gollasch S., Moore A., Tabe L.M., Craig S., Hardie D.C., Chrispeels M.J., Spencer D., Higgins T.J.V. (1995) Bean α-Amylase Inhibitor Confers Resistance to the Pea Weevil (*Bruchus pisorum*) in Transgenic Peas (*Pisum sativum*). Plant Physiol. 107: 1233–1239. http://www.plantphysiol.org/cgi/reprint/107/4/1233.pdf.

Snow A.A., Pilson D., Riesenberg L.H., Paulsen M.J., Pleskac N., Reagon M.R., Wolf D.E., Selbo S.M. (2003) A Bt transgene reduces herbivory and enhances fecundity in wild sunflowers. Ecological Applications 13(2):279–286, Ecological Society of America.

Spin Profiles (2010) Alan McHughen. http://www.spinprofiles.org/index.php/Alan_McHughen.

Spinprofiles.org. Florence Wambugu.
   http://www.spinprofiles.org/index.php/Florence_Wambugu.

University of Califormia, Riverside (2009) Alan Mc Hughen.
   http://www.plantbiology.ucr.edu/faculty/mchughen.html.

Warick J. (2001) GM Flax seed yanked off Canadian market – rounded up, crushed.
   http://www.rense.com/general11/gm.htm.

Williams N. (2010) One new potato. Current Biology, Volume 20, Issue 7, R301, 13 April 2010. doi:10.1016/j.cub.2010.03.040.
   http://download.cell.com/current-biology/pdf/PIIS0960982210003659.pdf?intermediate=true

Young E. (2005) GM pea cause allergic damage in mice. New Scientist.
   http://www.newscientist.com/article/dn8347-gm-pea-causes-allergic-damage-in-mice.html.

Theorie in der Ökologie

Herausgegeben von Broder Breckling

Band 1   Broder Breckling / Felix Müller (Hrsg.): Der Ökologische Risikobegriff. Beiträge zu einer Tagung des Arbeitskreises "Theorie" in der Gesellschaft für Ökologie vom 4.-6. März 1998 im Landeskulturzentrum Salzau. 2000.

Band 2   Kurt Jax (Hrsg.): Funktionsbegriff und Unsicherheit in der Ökologie. Beiträge zu einer Tagung des Arbeitskreises "Theorie" in der Gesellschaft für Ökologie vom 10. bis 12. März 1999 im Heinrich-Fabri-Institut der Universität Tübingen in Blaubeuren. 2000.

Band 3   Hauke Reuter: Individuum und Umwelt. Wechselwirkungen und Rückkopplungsprozesse in individuenbasierten tierökologischen Modellen. 2001.

Band 4   Fred Jopp / Gerd Weigmann (Hrsg.): Rolle und Bedeutung von Modellen für den ökologischen Erkenntnisprozeß. 2001.

Band 5   Kurt Jax: Die Einheiten der Ökologie. Analyse, Methodenentwicklung und Anwendung in Ökologie und Naturschutz. 2002.

Band 6   Franz Hölker (ed.): Scales, Hierarchies and Emergent Properties in Ecological Models. 2002.

Band 7   Achim Lotz / Johannes Gnädinger (Hrsg.): Wie kommt die Ökologie zu ihren Gegenständen? Gegenstandskonstitution und Modellierung in den ökologischen Wissenschaften. Beiträge zur Jahrestagung des Arbeitskreises Theorie in der Gesellschaft für Ökologie vom 21.-23. Februar 2001 im Kardinal-Döpfner-Haus Freising (Bayern). 2002.

Band 8   Katrin S. Romahn: Rationalität von Werturteilen im Naturschutz. 2003.

Band 9   Hauke Reuter / Broder Breckling / Arend Mittwollen (Hrsg.): Gene, Bits und Ökosysteme. Implikationen neuer Technologien für die ökologische Theorie. 2003.

Band 10 Thomas Potthast (Hrsg.): Ökologische Schäden. Begriffliche, methodologische und ethische Aspekte. 2004.

Band 11 Angela Weil: Das Modell „Organismus" in der Ökologie. Möglichkeiten und Grenzen der Beschreibung synökologischer Einheiten. 2005.

Band 12 Fred Jopp / Silvia Pieper (Hrsg.): Bodenzoologie und Ökologie. 30 Jahre Umweltforschung an der Freien Universität Berlin. 2008.

Band 13 Boris Schröder / Hauke Reuter / Björn Reineking (eds.): Multiple Scales in Ecology. 2007.

Band 14 Broder Breckling / Hauke Reuter / Richard Verhoeven (eds.): Implications of GM-Crop Cultivation at Large Spatial Scales. Proceedings of the GMLS-Conference 2008 in Bremen. 2008.

Band 15 Denis Worlanyo Aheto: Implication Analysis for Biotechnology Regulation and Management in Africa. Baseline Studies for Assessment of Potential Effects of Genetically Modified Maize (Zea mays L.) Cultivation in Ghanaian Agriculture. 2009.

Band 16 Broder Breckling / Richard Verhoeven (eds.): Large-area effects of GM-Crop Cultivation. Proceedings of the Second GMLS-Conference 2010 in Bremen. 2010.

www.peterlang.de